PROFESSIONAL DEVELOPMENT AND KNOWLEDGE OF MATHEMATICS TEACHERS

Mathematics teaching and professional development of mathematics teachers are areas where research has increased substantially in recent years. In this dynamic field, mathematics teaching practices, pedagogical knowledge of mathematics teachers and professional development via collaboration between mathematics teachers have emerged as vital domains of inquiry.

Professional Development and Knowledge of Mathematics Teachers addresses the underlying characteristics of mathematics teacher education, and those professional development contexts that have a positive impact on teachers' professional learning. Recognizing the impact of broader institutional settings on mathematics teaching and teacher professional development, the editors suggest bridging the gaps between theoretical practices and methodological approaches in the field by focusing on and conceptualizing the following relational factors:

- The study of mathematics teaching and classroom situations
- Researching teacher and teacher educator knowledge, since these issues inform the quality of mathematics teaching directly
- Mathematics teacher education and professional development, focusing on design principles and the impact they have on teacher professional learning

Combining central issues of mathematics teaching, knowledge and professional development, the chapters in this volume address each of the above factors and provide profound considerations on both theoretical and practical levels. This book will be an essential resource for researchers, teachers and students working in the fields of mathematics teaching and mathematics teacher professional development.

Stefan Zehetmeier is an associate professor at University of Klagenfurt, Austria. He holds a PhD in education and a habilitation in teacher education. His research interests include mathematics teacher education, school development and action research.

Despina Potari is a professor of mathematics education at the National and Kapodostrian University of Athens, Greece, and currently a visiting professor at Linnaeus University in Sweden. Her research interest is mainly on the development of mathematics teaching and teacher development.

Miguel Ribeiro is an associate professor at State University of Campinas – UNICAMP, Brazil. His research interests include teachers' knowledge and practices, teacher professional development and tasks for teacher education.

New Perspectives on Research in Mathematics Education – ERME series

Editors of the ERME Series:

Viviane Durand-Guerrier (France)
Konrad Krainer (Austria)
Susanne Prediger (Germany)
Nad'a Vondrová (Czech Republic)

International Advisory Board of the ERME Series:

Marcelo Borba (Brazil)
Fou-Lai Lin (Taiwan)
Merrilyn Goos (Australia and Ireland)
Barbara Jaworski (Europe, United Kingdom)
Chris Rasmussen (United States of America)
Anna Sierpinska (Canada)

ERME, the European Society for Research in Mathematics Education, is a growing society of about 900 researchers from all over Europe and beyond. In the ERME community and beyond, a growing body of substantial research on mathematics education has been raised which is shaped by the ERME spirit of communication, cooperation and collaboration.

The contributions in the ERME Series seek to understand and improve learning and teaching of mathematics at schools, colleges and universities, as well as in informal settings (e.g. related to street mathematics or to self-organized networks of teachers). The ERME Series puts an emphasis on reflecting the institutional, societal and cultural contexts of learners, teachers and researchers and how this context shapes research and adopts a variety of perspectives on its research field.

The volumes are written by and for European researchers, but also by and for researchers from all over the world. An international advisory board guarantees that ERME stays globally connected. A rigorous and constructive review procedure guarantees a high quality of the series.

Volumes of the series:

Working with the Anthropological Theory of the Didactic in Mathematics Education
A comprehensive casebook
Edited by Marianna Bosch, Yves Chevallard, F. Javier García, John Monaghan

Professional Development and Knowledge of Mathematics Teachers
Edited by Stefan Zehetmeier, Despina Potari and Miguel Ribeiro

For more information about this series, please visit: www.routledge.com/ European-Research-in-Mathematics-Education/book-series/ERME

PROFESSIONAL DEVELOPMENT AND KNOWLEDGE OF MATHEMATICS TEACHERS

Edited by Stefan Zehetmeier, Despina Potari and Miguel Ribeiro

Routledge
Taylor & Francis Group

LONDON AND NEW YORK

First published 2021
by Routledge
2 Park Square, Milton Park, Abingdon, Oxon OX14 4RN

and by Routledge
52 Vanderbilt Avenue, New York, NY 10017

Routledge is an imprint of the Taylor & Francis Group, an informa business

British Library Cataloguing-in-Publication Data
A catalogue record for this book is available from the British Library

Library of Congress Cataloging-in-Publication Data
Names: Zehetmeier, Stefan, editor. | Potari, Despina, editor. | Ribeiro, Miguel, editor.
Title: Professional development and knowledge of mathematics teachers / edited by Stefan Zehetmeier, Despina Potari & Miguel Ribeiro.
Description: Abingdon, Oxon ; New York, NY : Routledge, 2021. | Series: New perspectives on research in mathematics education-ERME series | Includes bibliographical references and index.
Identifiers: LCCN 2020034306 (print) | LCCN 2020034307 (ebook) | ISBN 9780367442408 (hardback) | ISBN 9780367442415 (paperback) | ISBN 9781003008460 (ebook)
Subjects: LCSH: Mathematics teachers—Training of. | Mathematics teachers—In-service training. | Mathematics teachers—Professional relationships. | Mathematics—Study and teaching.
Classification: LCC QA11.2 .P747 2021 (print) | LCC QA11.2 (ebook) | DDC 510.71/55—dc23
LC record available at https://lccn.loc.gov/2020034306
LC ebook record available at https://lccn.loc.gov/2020034307

ISBN: 978-0-367-44240-8 (hbk)
ISBN: 978-0-367-44241-5 (pbk)
ISBN: 978-1-003-00846-0 (ebk)

Typeset in Bembo
by Apex CoVantage, LLC

CONTENTS

FIGURES

TABLES

CONTRIBUTORS

Jill Adler, Wits School of Education, University of the Witwatersrand, South Africa

Edelmira Badillo, Autonomous University of Barcelona, Spain

Bärbel Barzel, University of Duisburg-Essen, Germany

Rolf Biehler, University of Paderborn, Paderborn, Germany

Laurinda Brown, University of Bristol, UK

José Carrillo, University of Huelva, Spain

Brent Davis, University of Calgary, Canada

Genaro de Gamboa, Autonomous University of Barcelona, Spain

Ronnie Karsenty, Weizmann Institute of Science, Israel.

Dorte Moeskær Larsen, University College Lillebælt, Denmark

Miguel Montes, University of Huelva, Spain

Diana Vasco Mora, Quevedo State Technical University, Ecuador

Camilla Hellsten Østergaard, University College Copenhagen and University of Copenhagen, Denmark

Despina Potari, National and Kapodistrian University of Athens, Greece

Susanne Prediger, TU Dortmund University, Germany

Miguel Ribeiro, University of Campinas, Brazil

Nuria Climent Rodríguez, University of Huelva, Spain

Kim-Alexandra Rösike, TU Dortmund University, Germany

Gloria Sánchez-Matamoros, University of Seville, Spain

Susanne Schnell, Goethe University, Frankfurt, Germany

Jeppe Skott, Linnaeus University, Sweden

Hamsa Venkat, Wits School of Education, University of the Witwatersrand, South Africa

Stefan Zehetmeier, University of Klagenfurt, Austria

1

MATHEMATICS TEACHING AND MATHEMATICS TEACHER PROFESSIONAL DEVELOPMENT

Stefan Zehetmeier, Despina Potari & Miguel Ribeiro

Mathematics teaching and mathematics teacher professional development are areas where research has increased substantially in recent years. In this ongoing field of research, many issues need further investigation. We need to better understand the underlying characteristics of mathematics teacher education and the professional development contexts that have a positive impact on teachers' professional learning. This includes in particular research about mathematics teaching practices and mathematics teacher knowledge, as well as about teacher professional development and collaboration. Moreover, research about mathematics teaching and mathematics teacher professional development relates to the classroom level as well as to the school and the broader institutional settings within an educational system. Research-informed professional development structures for teachers and teacher educators, especially at large scale, seem to be crucial for the improvement of mathematics teaching at school. In sum, further discussion is needed on how to connect particular research findings and how to link research with mathematics teacher education and classroom practices.

Researching this bridge between theory and practice encompasses a complex range of possible topics. To reduce this complexity, we may follow a path with several concrete stations, all of which are central to mathematics teaching and mathematics teacher professional development. Thus, this path can serve as a guiding line along which research may be conceptualized and located.

This path may start from the study of mathematics teaching and classroom situations as a first station (1). The path's stations may continue with researching teacher and teacher educator knowledge as a second station (2), since these issues inform the quality of mathematics teaching directly. Following this path may lead to a third station (3) regarding research on mathematics teacher education and professional development, focusing on design principles and the impact they have on teacher professional learning.

This book provides theoretical and empirical studies which follow this path of research. The chapters provide profound considerations on both theoretical and practical levels regarding each of these concrete stations: (1) mathematics teaching and classroom situations; (2) mathematics teacher and mathematics teacher educator knowledge and (3) mathematics teacher education and professional development. For each station, this book provides three chapters: the first chapter provides theoretical frameworks which are suited to analyze the content of this station, while the other two present research projects going deep into the respective realm. In sum, this book provides diverse perspectives and approaches which contribute to our knowledge and understanding about mathematics teaching and mathematics teacher professional development.

1. Research on mathematics teaching and classroom situations

The first chapter written by *Hamsa Venkat* and *Jill Adler*, "Mediating mathematics in instruction: Trajectories towards generality in 'traditional' teaching", provides two theoretical frameworks (the Mediating Primary Mathematics – MPM – framework at the primary level, and the Mathematical Discourse in Instruction – MDI – framework at the secondary level) which have been developed in the course of researching and developing mathematics teaching in primary and secondary schools in South Africa. In their chapter, they highlight the wide applicability of mathematical tools and results across various examples and situations. In particular, they illustrate through these frameworks the meaning of teaching for generality in the context of "traditional" instruction, where learners often imitate taught procedures. Finally, they discuss the rationales underlying the similarities and differences in the two formulations.

The second chapter, "The role of teachers' knowledge in the use of learning opportunities triggered by mathematical connections", focuses on the relationship between extra- and intra-mathematical connections and the role of a teacher's knowledge in the use of learning opportunities. *Genaro de Gamboa, Edelmira Badillo, Miguel Ribeiro, Miguel Montes* and *Gloria Sánchez-Matamoros* provide insights into empirical research on non-standard measurement in second grade of primary school and the connections related to the mathematical foundations of length measurement. Learning opportunities stemming from such connections are described and analyzed. In this chapter, the authors highlight that extra-mathematical connections are strongly based on intra-mathematical connections. In particular, different types of knowledge can help teachers to make the most of the learning opportunities arising from such connections.

Jeppe Skott, Dorte Moeskær Larsen and *Camilla Hellsten Østergaard* in their chapter, "Learning to teach to reason: Reasoning and proving in mathematics teacher education", provide an empirical intervention study which addresses the problems of reasoning and proving in mathematics teacher education in Denmark. School mathematics and teacher education aim at emphasizing proving "why" rather than proving "that" when teaching reasoning and proving in schools. From this background, authors outline the background, framework and results of their empirical study. In particular, they highlight that teachers face problems with reasoning and proving and have difficulties selecting adequate classroom situations. The authors suggest a dual emphasis on both proving that and proving why in mathematics teacher education.

2. Research on teacher and teacher educator knowledge

In her theoretical chapter, "The role of frameworks in researching knowledge and practices of mathematics teachers and teacher educators", *Ronnie Karsenty* discusses the notion of "framework" and its utilization for conceptualizing knowledge and practices of mathematics teachers and mathematics teacher educators. She focuses on double-level use of frameworks, referring to frameworks that can serve for the purpose of researching both teachers and teacher educators. This chapter explores in particular if and how frameworks from classroom level can be transferred to the level of professional development. Moreover, a specific case of framework adaptation from teacher level to facilitator level is presented.

The second chapter of station (2), "Parallel stories: teachers and researchers searching for mathematics teachers' specialized knowledge", written by *José Carrillo*, focuses on the relationship between mathematics teachers and mathematics education research. It describes the joint learning trajectory, which develops when working in the teacher professional development realm. A particular focus is on teacher knowledge as a central fostering factor promoting professional development. The author reviews the Mathematics Teacher's Specialized Knowledge (MTSK) model and gives an empirical example of using this model when analysing one particular teacher's classroom activities in Spain.

The following empirical chapter of *Diana Vasco Mora* and *Nuria Climent Rodríguez*, "The specialized knowledge and beliefs of two university lecturers in linear algebra", uses the aforementioned MTSK model to analyse university mathematics lecturers' knowledge and beliefs in teaching linear algebra. In particular, the chapter focuses on the use of knowledge and beliefs as personal resources. Using qualitative research and an instrumental case study design, the authors found evidence that elements of content knowledge, pedagogical content knowledge and beliefs appear to be consistent with each other. The study's results point to the central role of examples used by the lectures when supporting students to overcome mathematical difficulties.

3. Research on mathematics teacher education and professional development

Laurinda Brown and *Brent Davis* in the first chapter of this station, "Using the discourses of learning in education mapping to analyse research into mathematics teacher education and professional development", deal with the question of theoretical perspectives on learning. To do this, they use a theoretical framework for analysing mathematics teacher education and professional development: the Discourses on Learning in Education. In particular, they use two case studies (concerning Variation Theory and Professional Learning Community) to illustrate their analysis. The authors argue that any theoretical or analytical framework for analysing mathematics teacher education and professional development should be based on images and metaphors used by researchers and theorists to characterize learning and learners. Moreover, they suggest that such a framework should be more about description than prescription, to be able to expand understanding and communication.

Content-specific Design Research is provided by the chapter "Promoting and investigating teachers' professionalization processes towards noticing and fostering students' potentials: A case of content-specific Design Research for teachers" written by *Susanne Prediger, Susanne Schnell* and *Kim-Alexandra Rösike*. Against the backdrop of a specific teacher professional development content (called "noticing and fostering students' mathematical potentials"), they present the conduction and analysis of design experiments. Particular focus is on the specification of what teachers have to learn and the respective design of professional development. Moreover, exemplary outcomes regarding teachers' noticing and fostering students' potentials are discussed.

Bärbel Barzel and *Rolf Biehler*, in the chapter "Theory-based design of professional development for upper secondary teachers – focusing on the content specific use of digital tools", provide research results regarding the effects of professional development programmes on teachers' beliefs and knowledge in Germany. One exemplary programme focuses on digital tools fostering process-related competences (such as modelling, flexible use of representations and problem-solving), another programme is related to probability and statistics as content. In particular, these programmes aim at supporting teachers in implementing process and content-related competences and digital tools in their classrooms.

2

MEDIATING MATHEMATICS IN INSTRUCTION

Trajectories towards generality in "traditional" teaching

Hamsa Venkat & Jill Adler

1. Introduction

Ole Skovsmose's (2011, p. 18) conjecture that "90% of research in mathematics education concentrates on the 10% of the most affluent classroom environments in the world" has been important to our research and development work in South Africa. Skovsmose outlines the implications of this skew in the empirical sites of mathematics education research for the theoretical frameworks available for take-up in the field. Specifically, he notes the lack of need to deal with issues of widespread poverty, unemployment and hunger, with consequences in terms of developmental delays, of classrooms with limited access to electricity and few educational resources, of large classes and overcrowded classrooms and of authoritarian instructional forms. Some combinations of these issues continue to be realities in many government schools in post-apartheid South Africa, in spite of increased expenditure on education and successes relating to getting near universal access to primary education and improvements in provision of basic resources such as national workbooks.

In supporting and studying classrooms in these realities, we share the commitment to expand access to richer and more connected mathematical experiences that underpins much of the research in Skovsmose's dominant contexts. Our own experiences and broader research in South African schools paint a picture of mathematics classrooms with a different "base" from these dominant contexts that our research and development efforts have had to take into account. In this chapter, our focus is on frameworks we have developed for analyzing and supporting mathematics teaching that have sought this contextual salience while retaining aspirations for high-quality mathematics teaching that would be recognizable in the international research base. While the frameworks differ across primary and secondary mathematics, they share common bases in socio-cultural theory. Key tenets in focus are a view of instruction as mediating learning via a range of mediating means, and as goal-directed activity towards learning mathematics as a network of scientific concepts, in which generality and increasingly complex disciplinary thinking are sought. These tenets are broadly shared in the international mathematics education community. Our emphasis on the teacher's handling of mathematics within the two frameworks places our work at the content-specific end of Charalambous and Praetorius (2018) generic- to content-specific continuum and more instruction- than interaction-facing than the frameworks that have received wide attention in better-resourced contexts. The latter aspect reflects classroom cultures in which teacher-directed whole-class instruction remains highly prevalent.

Our focus in this chapter is on the different ways in which we have considered mathematical generality across the two frameworks (the Mediating Primary Mathematics (MPM) framework at the primary level, and the Mathematical Discourse in Instruction (MDI) framework at the secondary level). In the ERME Topic Conference paper that was the precursor to this chapter, our focus was on the analysis of instructional talk as a key mediating form across these two frameworks, and how quality towards generality was configured as a goal within this strand (Venkat & Adler, 2017). In this chapter, we home in on the notion of generality more specifically and consider the ways in which generality is considered in relation to teachers' work across mediating forms in the MPM and MDI models within tasks/examples and their broader instructional discourse. While the frameworks themselves and the aspects in focus within them have been written about elsewhere (Venkat & Askew, 2017, 2018; Adler & Ronda, 2015, 2017), our attention here is on the theoretical strands in the international literature we have drawn from to consider generality, across primary and secondary levels, with emphasis on continuities and adaptations to existing theoretical formulations based on an anchoring in a ground of largely traditional instructional forms. In the South African context, "traditional" forms commonly include a teacher-directed "transmissionist" form of instruction, with chorused responses to mainly routine problems. In this chapter, we explore overlaps and differences in the theoretical derivations that underlie the ways in which we work with generality.

Generality can frequently feel like a distant goal in the South African context, given the evidence of low performance in mathematics at all levels of the schooling system. This raises questions about why we chose to develop frameworks oriented towards generality as a key goal. Our response would be that in a South African terrain where we have highlighted frequent disconnection and incoherence in mathematics instruction (Venkat & Adler, 2012; Adler & Venkat, 2014), this theoretical orientation provided a useful and important counterpoint and aspiration that could guide and focus our research and professional development activity. But a goal of generality in isolation of a trajectory for its achievement was likely to lead us down a rabbit hole of deficit analyses that would defeat our development objectives. Thus, across the two frameworks, our attention was on trajectories towards generality that could inform our teacher development activity as well as our research analyses and evaluations of instruction. We begin this chapter with a short overview of generality as conceived in socio-cultural theory and Vygotsky's work on scientific concepts in particular. Brief overviews of the MPM and MDI frameworks are then provided, prior to our presentation of the ways in which coding for generality is configured within the two models drawing on Variation Theory with distinct and yet overlapping approaches. This presentation leads into a discussion of the overlaps and contrasts between the primary and secondary mathematics terrains in South Africa, and our differing theoretical histories, that feed into the differing configurations of generality in the frameworks. We conclude with a return to Skovsmose's (2011) critique and note the ways in which our work might feed into broadening the scope of the contexts that feature in the international mathematics education research terrain.

2. Generality

A central tenet of Vygotsky's (1987) notion of learning was that it was geared towards the development of what he termed as higher psychological functions. For Vygotsky, teaching for the development of higher psychological functions depended on mediation of disciplines viewed as networks of related and connected "scientific concepts". Kozulin (2003), expanding on Vygotsky's ideas, described mediation that pulled towards increasing generality or towards the more specialized ways of thinking that define disciplines as moves towards the goal of scientific, rather than spontaneous, concepts. Karpov and Bransford (1995) emphasize that instruction is central to this endeavour, pointing out that while children may well generalize from their everyday experiences, the danger inherent in this kind of spontaneous generalization is that the concepts formed could remain at the level of the "unsystematic, empirical and unconscious" (p. 61), without coalescing into the organized and interconnected networks that constitute disciplines.

Watson and Mason's extensive body of work in mathematics sees generality as produced through seeing invariance (the commonality) across variation in relation to examples and example spaces. This approach was useful to us, given the familiarity of tasks and examples in traditional teaching contexts.

As noted earlier, the focus on generality provided a counterpoint to the issue of "localization" which has been written about extensively in South Africa and elsewhere from a range of different theoretical perspectives. In sociological writing (e.g. Hoadley, 2007), localization is contrasted with specialization and draws from a base in which informal, or everyday knowledge (usually tied to a specific contextual base in personal experience) is contrasted with formal knowledge (knowledge that has generality across contexts). Formal knowledge gains its power specifically by being delinked from everyday, or contextual familiarity.

In Sfard's (2008) approach to thinking as communication and mathematics as a discourse, generality is related to mediation towards the symbolic, and contrasted with both concrete and visual mediators, whose reach, while important for communication, is small (i.e. situated and thus localized):

> The special strength of iconic and concrete mediators is that they may lead to new endorsed narratives with only a relatively small number of verbal manipulations (reasoning actions). The symbolic means, on the other hand, are basically verbal and thus sequential and as such exert greater demands on one's memory. And yet, what is lost in simplicity is gained in generalizability and applicability.
>
> *(p. 162)*

Watson and Mason's writing, with a base in Variation Theory, links localized working to finding "immediate answers" involving repeated work with separate examples. In our earlier work, we have incorporated this latter line, looking at localization particularly in terms of working with examples in highly "separate" ways (Venkat & Naidoo, 2012; Adler & Venkat, 2014).

From a Variation Theory perspective, variation across examples is viewed as a necessary condition for the possibility of noticing distinctions, and generality is seen as a potential outcome of using example spaces to abstract invariance in the midst of variation (Marton & Booth, 1997). Across the MPM and MDI frameworks, generality is approached in relation to example spaces in different ways. In the MPM, generality is approached through considering Watson and Mason's body of writing on using variation/invariance in example spaces. In their work, they use the term "dimensions of possible variation" to refer to the aspects that can be varied (quantities, representations of quantities and relations among these) and "range of permissible change" to refer to the range of types or values that each aspect can take (Watson & Mason, 2006). Increasing generality refers to expansions in the example space that can be worked with. In the MDI model, generality is approached using and adapting constructs from Swedish work on variation (e.g. Marton & Tsui, 2004), specifically the constructs of similarity, contrast and fusion. Focusing on what something is through a set of *similar examples* (with a key feature invariant while others vary) brings possibilities for generalizing that which is invariant. Similarity on its own, however, does not draw attention to the boundaries around an object, and so to what it is not. *Contrasting examples* that bring attention to a different class also make opportunity for generality available. We also identify when there is *fusion* with more than one aspect of an object of learning varying/invariant simultaneously across an example set. Further details on both of these approaches are provided in the sections that follow.

3. Introducing the frameworks

The MPM and MDI frameworks commonly draw from the key tenets of sociocultural theory that we outlined earlier in this chapter: mathematics viewed as an interconnected network of scientific concepts, and mathematics instruction therefore geared towards appropriation of the increasingly sophisticated and increasingly general ways of thinking that constitute progression in the discipline. As academics, we both come from research backgrounds in which activities such as teaching are seen as "social" in their base. In the first author's history, this background has drawn from activity theoretical strands of socio-cultural theory in which concepts such as mediational forms and goal-oriented activities have been of interest (e.g. Venkatakrishnan, 2005); in the second author's history, the background draws from theoretical strands in socio-cultural and sociological theory in which concepts such as mediational means, particularly language, are understood as cultural tools or resources in a practice (Adler, 2001); and legitimate text as that which comes to count as mathematics in school lessons (e.g. Adler & Davis, 2006).

3.1 The MPM framework

The MPM framework was devised as a tool to analyse, support and evaluate primary mathematics instruction. The model considers instruction in mediational terms, comprised of a semiotic bundle (Arzarello, 2006) that incorporates tasks/examples, artefacts, inscriptions and instructional talk. This approach emanated from instances in our baseline observations and other research of incoherence in the orchestration of elements of the semiotic bundle (e.g. talk that did not cohere well with the artefact being used or the example in focus) (Mathews, Venkat & Askew, 2018), and examples being worked with in highly separate, rather than connected, ways (Venkat & Naidoo, 2012).

Within the MPM analytic approach, observed lessons are chunked into episodes based on the teacher's work with a particular mathematical idea. The mathematical idea in focus has tasks/examples associated with it that form the raw material upon which mediation via artefacts, inscriptions and teacher talk is enacted. In our coding of episodes, we therefore begin with the identification of the mathematical idea being worked with (sometimes explicitly stated, and at other times inferred from the tasks/examples and mediation) and list the tasks/examples associated with this idea. The ways in which these tasks/examples combinations are mediated are coded across the semiotic bundle strands of artefacts, inscriptions and talk. The talk strand in the MPM model encompasses gesture and is broken down into three sub-strands:

- Talk/gesture in relation to generating or validating methods for producing solutions to problems
- Talk/gesture in relation to building mathematical connections
- Talk/gesture in relation to building responsive learning connections

Across all strands, coding works from a base in error/incoherence into coherence, then connection and finally into mathematical structure and generality. The overall framework is outlined in Figure 2.1 and includes some exemplifications,

MPM Framework — **MEDIATING TASKS/ EXAMPLES** — Venkat&Askew, 2018

MEDIATING ARTEFACTS

0	1	2	3
No artefacts or artefacts that are problematic/inappropriate	Unstructured artefacts used in unstructured ways (Bags of counters/tally marks)	Structured artefacts used in unstructured ways (Abaci, 100 squares, etc., used with unit counting, and without reference to structural properties)	Structured artefacts used in structured ways/unstructured artefacts used in structured ways (Abacus, 100 square/place value blocks/cards, number lines, etc., used with reference to structure/relations)

MEDIATING INSCRIPTIONS

0	1	2	3
No inscriptions or inscriptions that are problematic/incorrect	Inscriptions that record only tasks or responses	Unstructured inscriptions (e.g. tally marks)	Structured inscriptions (e.g. tables of ordered bonds; structured/empty number lines, crazy grids; inscriptions underpinned by number relations)

MEDIATING TALK & GESTURE

	0	1	2	3
Method for generating/ validating solutions	No method or problematic generation/validation (e.g. mixing of knowns and unknowns)	Singular method/validation (provides a method that generates the immediate answer; enables production of answers in the immediate example space)	Localized method/validation (provides a method that can generate answers beyond the particular example space)	Generalized method/validation (provides a strategy/method that can be generalized to both other example spaces AND without restriction to a particular artefact/inscription)
Building mathematical connections	Disconnected and/or incoherent treatment of examples OR oral recitation with no additional teacher talk	Every example treated from scratch	Teacher talk connects between examples or artefacts/inscriptions or episodes	Teacher talk makes vertical and horizontal (or multiple) connections between examples/ artefacts/inscriptions/episodes
Building learning connections: explanations and evaluations – of errors/for efficiency/with rationales for choices	Pull-back to naive methods OR No evaluation of offers (correct or incorrect)	Accepts/evaluates offers Accepts a strategy or offers a strategy OR Notes or questions incorrect offer	Advances or verifies offers Builds on, acknowledges or offers a more sophisticated strategy OR addresses errors/misconceptions through some elaboration, e.g. 'Can it be —?' 'Would – this be correct, or this?' Non-example offers	Advances **and** explains offers Explains strategic choices for efficiency moves OR provides rationales in response to learner offers related to common misconceptions OR provides rationale in anticipation of a common misconception

FIGURE 2.1 The MPM framework.

with detail on its concepts and theoretical background in Venkat and Askew (2018).

In our evaluations of the quality of lessons, we developed the additional concepts of "extent" and "depth" of mediation for structure and generality (Askew et al., 2019). For this, we combined artefacts and inscriptions into a single combined strand as the relatively "permanent" mediation forms separate from the talk/gesture strands as the more ephemeral and potentially responsive mediating forms, creating four mediating strands. The extent indicator involves a "horizontal" summing of the scores out of the total possible score for each mediating form across coded episodes in lessons, producing a proportion that can be used for comparing teaching over time or for comparing between lessons. The depth indicator involves reading the MPM coding of episodes in a lesson. Here we consider the proportion of episodes in a lesson that contain higher-level (2 or 3) codes across two or more strands. The depth indicator, too, produces a proportion that can be used for comparing teaching over time or for comparing between lessons (see Askew et al., 2019, for detail on the coding approach).

3.2 The MDI framework

The MDI framework has both descriptive/analytic and evaluative components emerging from our intertwined research interests in describing the mathematical made available to learn in secondary school mathematics lessons and evaluating the quality of mathematics in instruction in and over time. Our emphasis on *mathematics in instruction* is apposite, as is our focus on a lesson as a key unit of teaching in secondary schools in South Africa. Typically, these range from 30 to 45 minutes and have a particular content in focus. While this cultural form is not unique to South Africa (e.g. Alexander, 2000), it is aligned with the National Curriculum that recommends X number of weeks of instruction for a particular topic. In addition, at the provincial level, there is an Annual Teaching Plan, elaborating what is to be done in the X weeks in order to support "coverage" of the curriculum. In our initial observations of numerous mathematics lessons in the secondary schools in our project, and notwithstanding the levels of prescription described earlier, we struggled to interpret the intended mathematical message in a lesson and wondered how specific mathematical goals prefigured lesson development activity for teachers. We described this phenomenon as lessons in which, despite an announced topic or content, the mathematical object of learning was out of focus. For example, in a four-part lesson ostensibly on multiplying algebraic expressions, different rules reliant on visible forms of the expressions were emphasized in each part, making available a fragmented and incoherent notion of the products (Adler & Venkat, 2014). MDI emerges in response to this observation, as we worked to devise an analytic tool that could support our research and development work with mathematics teachers.

Our starting point in MDI is that teaching, and thus learning, is always about *something*, and bringing that into focus – its mediation – is the teacher's work

(Adler & Ronda, 2015). This "something", using Marton and Tsui's (2004) language, is the *object of learning*, consisting of the mathematical "object" and the capability associated with it that the teacher aims to focus on in the lesson. While in practical terms, it aligns with a lesson goal, it points to that which needs to be mediated and thus the goal-directed activity of the teacher. In the MDI framework captured in Figure 2.2, the key generative mechanisms for the work of teaching are exemplification, explanatory communication and learner participation (for detail see Adler & Ronda, 2015). What stands between (i.e. mediates) the object (and here of learning) and the subject (the learner) are four key symbolic mediational means in mathematics classroom instruction – *examples, tasks, names and legitimations* – that work together with learner participation with these means, to exemplify and elaborate/explain the object of learning – what it is about and what learners are to be able to do with/on/in relation to it. As illustrated in Figure 2.2, all mediation is towards the building of scientific concepts, and so towards increasing generality or towards the more specialized ways of thinking and working mathematically.

Each of the four instructional tools (semiotic in nature and thus, following Venkat and Askew [*op. cit.*], can also be described as semiotic bundles) has theoretical antecedents in the literature on research in mathematics education, as elaborated later in this chapter, and we use these separately and in combination to describe the mathematics made available to learn in a lesson. As we are concerned with mathematical coherence in relation to an intended object of learning, we chunk the lesson into *mathematical episodes* that are identified by a specific focus of attention with respect to mathematics content, typically marked by a task that encompasses selected example(s) and bears some relation to the stated object of learning, and a next episode when there is a move to another specific focus.

We examine each episode for its exemplification (its examples and tasks), explanatory communication (how objects are named, and actions on these legitimated) and how learners are invited to participate in their learning. The detail of the coding for this analysis is in Adler and Ronda (2015). Following the coding

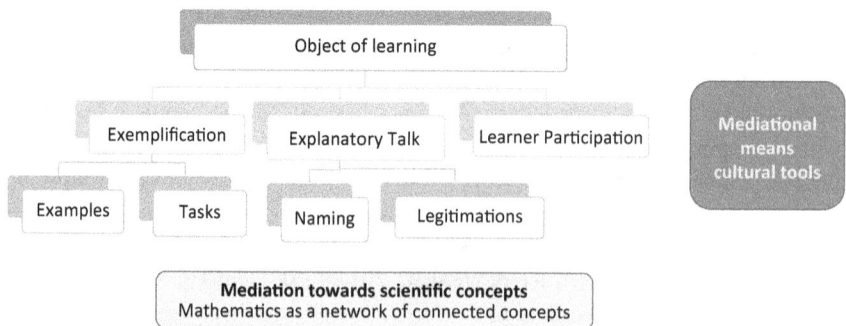

FIGURE 2.2   Constitutive elements of MDI.

Source: Adapted from Adler and Ronda (2015).

within each episode, we look across the lesson to evaluate the cumulative possibilities opened up (or not) to experience rich mathematics and high-quality mathematics teaching. This evaluative part of the MDI framework is presented here, following further discussion of how we view the building of generality within each of the four elements of MDI. The mathematical episodes also combine to build the mathematical story in the lesson, and thus they provide a window into coherence. The overarching importance of the framework is the emphasis on the coherence of *a lesson* and thus how all elements interact, link back to the object of learning and open opportunities to learn mathematics.

The presentation of the frameworks themselves brings some common aspects and distinctions into view. We return to discuss distinctions relating to the ways in which we work with tasks/examples later in the chapter, following a discussion of how generality is conceptualized across MPM and MDI.

4. Generality as a goal in MPM and MDI

We have already pointed to generality as a goal in our underlying view of mathematics as a scientific discipline. This goal guides the ways in which we consider and code instruction across the MPM and MDI, drawing from some overlapping and some distinct theoretical strands in our approaches. The methodological approaches and the theoretical bases underlying these approaches for the MPM and MDI are detailed in the sections that follow.

4.1 Generality in the MPM

In the MPM framework, the formulation of each strand is guided by a trajectory towards generality from a ground characterized by fragmentation. Across artefacts and inscriptions, teachers' work with these mediating forms in ways that are underpinned by attention to mathematical relationships, forms the key indicator for instruction marked by attention to structure. In working with relationships between entities (particularly relationships between numerical quantities given our location in primary schooling where number forms the largest topic area) rather than with entities as separate, coherent compositions of connections firstly within examples and then between examples lay the ground for generality to be brought into view. Kieran's (2018) writing on structure as involving de- and re-compositions of mathematical objects into mathematically appropriate relationships is a key underpinning idea here.

Generality is incorporated in different ways into the three sub-strands of talk. In the MPM model, tasks and examples are listed within episodes across lessons. Following Watson and Mason (2006), tasks and examples are seen as the "raw material" upon which mediation through artefacts, inscriptions and talk is enacted. Drawing on the theoretical base in Variation Theory, taken up via Watson and Mason's work (2006) on example spaces, we are interested in looking at the mediation of the example space in the semiotic bundle. Specifically, we pay attention to how artefacts, inscriptions and talk/gesture are assembled in ways that allow learners to function, first, within the example space provided, and then beyond this example space.

This approach is based on evidence of the usefulness of considering the ways in which instruction supports independent learner working within the terrain of the kinds of examples presented, and the "reach" of instruction beyond the immediate example space and into related sets of examples. The notion of "dimensions of possible variation" provides a handle for thinking about the aspects that can be varied in a given example space to produce related examples – e.g. the quantities in the problem, or the artefact or inscription used to represent the problem situation, or the solution procedure. For instance, where a set of examples (all numbers less than 90) were worked with as a base for adding 10, we noted instruction where this idea was worked with using a 100 square and teacher talk/gesture entrenched the idea of "jumping down one square". While the artefact and instructional talk cohered with each other here, we noted that no openings were provided for learners to function beyond the localization in the specific artefact provided, nor for moving beyond the number range in the 1–100 grid. Thus, in this kind of case, there was a coherent mediation of the examples, but a mediation that remained localized within the artefact provided by the instructional talk offered and within the number range of the examples presented. While the procedure offered in this case ('jump down a square') had potential to go beyond the 1–100 number range, an explicit mediation for this transcending would have required extensions in the examples provided, and/or in the artefacts presented and/or in the marshalling of connections through the use of inscriptions and/or talk. In the generating/validating procedures substrand of talk, this mediating talk would therefore be coded at Level 1.

In this consideration of the reach of the procedure offered, we draw on and adapt Watson and Mason's (2006) notion of "range of permissible change". In their work, Watson and Mason study the range of values possible within some dimension of variation that allows a stated condition to be met. In their "taxi-cab geometry" task in the aforementioned paper, a carefully sequenced set of examples are presented that support the making, and re-making, of generalizations about the range of coordinates that are a distance of 3 units away from a given fixed point, if distance is measured as a combination of horizontal and vertical moves.

In this formulation, range of permissible change provides a route into the creative development of example spaces to encourage generalizing activity and increasingly complete generalizations. While we agree with this formulation, the primary mathematics curricular terrain in South Africa is characterized by a prescriptive curriculum that stipulates topic selection, sequencing and pacing, with increasing provision of associated lesson plans. This specification follows prior evidence of poor pacing and progression (Reeves & Muller, 2005), with the latter seen in the ongoing use of highly time-consuming unit-counting strategies for solving arithmetic problems. Therefore, the cultural terrain and the policy terrain in terms of artefact provision in the form of lesson plans and workbook schedules of coverage tend to work against an orientation geared towards the deliberate selection of examples.

The literature base on the role of examples in mathematics suggests the usefulness of careful selection and sequencing of examples (e.g. Zodik & Zaslavsky, 2008). Skilful development of early mathematical learning is also allied with access to, and then the fading of concrete artefacts and actions as they give way to increasingly truncated symbolic inscriptions (Gray, 2008). Alongside this orchestration, Treffers (2001) points to strategic progression in the procedures used to solve problems from calculating by counting, calculating by structuring, into pure calculating.

Given the evidence of key gaps in progression relating to the sophistication and generality of procedures, and given, too, the familiarity with teaching procedures in traditional mathematics classrooms, we chose to adapt Watson and Mason's (2006) approach to working with the notion of "range of permissible change". Rather than looking at the range assigned to variables within particular dimensions of variation in the example space, we focus instead on the procedures offered or accepted within episodes for solving problems within teachers' explanations, and we analyse these procedures in terms of how far the example space can be "stretched" with the procedure in focus. In the instance discussed earlier, the procedure based on "jump down one square" has some generality in its application beyond the example space provided, but is localized in the instruction to the physical presence and use of the 1–100 square, with no attempt to fade the artefact as a way to transcend this localization.

Generality in the first sub-strand of talk is therefore considered from the vantage point of the potential of the procedure presented, rather than in relation to the potential of the example space. In this way of thinking about the Variation Theory concepts of dimensions of possible variation and range of permissible change, there are overlaps with the writing on variation from China where emphasis on the teaching and learning of increasingly powerful and general procedures has been widely discussed (Gu, Huang & Gu, 2017). In particular, our approach overlaps with what Cai and Nie (2007) have outlined as the "multiple problems, one solution" approach, in which emphasis is on the generalizability of a problem-solving method or procedure across a class of examples.

In the second sub-strand of talk, generality is produced via the building of increasingly extensive mathematical connections in the example spaces being worked with. Beginning from a base in incoherent, or problematic working with examples (level 0), moves towards improving quality involve coherent but separate treatment of individual examples such as that described in Venkat and Naidoo (2012) and Adler and Venkat (2014) at Level 1, and then connections between episodes, examples or artefacts/inscriptions at Level 2 and multidimensional connections across mediating forms for Level 3. Through these connections made within and between examples, and then between episodes, examples start to be treated as a connected "set" with the possibilities for attention to structure and generality entailed in this kind of mathematical connection in instruction (Askew, 2019).

In the last sub-strand of talk, generality is engendered through attention to the extent to which instruction can be seen as "responsively" appropriate. Leinhardt's (1990) writing on instructional explanations forms a key base for the trajectory delineated in this sub-strand. Generality here is built through attention to remediating in the context of incorrect learner offers and advancing in the context of correct offers. Given that remediation or advancing learner offers can be seen as central to responsive teaching, part of the attention in this strand is drawing attention to multiple methods of solving problems (Cai & Nie, 2007), with particular sensitivity to progression towards more efficient (and thence more structural) approaches over time with "unpacking" where needed. Generality can also be supported through the offer of rationales for choices or actions that provide openings for the independent construction of problem-solving processes by learners. In a context of evidence of limited responsive evaluation, and sometimes an absence of evaluation of any sort of learner offers (Hoadley, 2007), the trajectory towards generality in this strand works from a base in the absence of evaluation, towards limited and then more extensively responsive evaluation that builds in rationales that support the kinds of independent learner working on problems that underpin generalized competence.

4.2 Generality in the MDI

In MDI, with respect to *examples*, we are concerned with what *can* be brought into focus through an *example set* across an entire lesson, and what *actions* (*tasks*) learners are invited to carry out on the examples. Different theoretical strands inform our work with examples and tasks. We draw on Watson and Mason's (2006) attention to variance amidst invariance to examine what features of the object of learning are varying while others are kept invariant. Our notion of example set is similar to what Goldenberg and Mason (2008) call an *instructional example space*. As discussed earlier in this chapter, we draw on the constructs of similarity, contrast and fusion in Swedish Variation Theory (e.g. Marton & Tsui, 2004), so as to examine: similarity across the example set, and thus possibilities for generalizing that which is invariant; contrasting examples in the set that bring attention to the boundary of the generality; whether and how more than one aspect of an object of learning are varying/invariant simultaneously (fusion) across an example set. While we categorize the examples within a mathematical episode, coding for these features of variation, we evaluate the mathematical quality of the example set by looking across episodes and thus possibilities through similarity, contrast and fusion across the whole example set to open possibilities for movement towards generality, agreeing with Mason, Graham and Johnston-Wilder (2005) that possibilities for generalizing mark out mathematical quality in a lesson.

a lesson without an opportunity for learners to generalize cannot be considered to be a mathematics lesson. In other words, in every topic, in every exercise, in every task, there are opportunities for generalisation.

(p. 323)

For example, in the lesson on multiplying algebraic expressions mentioned earlier, the discerning of the general rule for multiplying a polynomial by a monomial (each term in the polynomial must be multiplied by the monomial) was made possible in the example set as the teacher proceeded to expand $4(x + 2) =$, $4x(x + 2) =$ and $-4x(x + 2) =$ followed by $2x(3x^2 + 2x - 4) =$. The binomial term was kept invariant to start, varying the monomial term and ultimately both the monomial and the polynomial, thus with attention to similarity. Contrast emerged with the fifth example, $(x + 2)(x + 3) =$, drawing attention to the limits of the generality to this point, as the multiplier was now a binomial. Following the elaboration of this new product, a general rule for which would at this point be limited (e.g. the operations/signs are all +), a classwork "exercise" was set, including a range of products of binomials, and extended to expressions with the addition of products e.g. $2(y - 3) + y(y - 4) =$.

Our judgement of an example set such as this, and how it accumulates across the lesson, would be high (level 3 in our evaluative part of the framework), as through similarity, contrast and fusion, there is possibility for movement towards generality and for extending disciplinary thinking.

In MDI we define a *task* as simply what learners are asked to do with the various examples presented. Thus, while examples are selected as "particular instances" of the general case in focus, and for drawing attention to "relevant features", tasks are designed to bring particular capabilities to the fore (Marton & Pang, 2006; Marton & Tsui, 2004). Notwithstanding the abundance of recent literature and research on task design in mathematics education, we confine ourselves to the distinction made by Stein, Smith, Henningsen and Silver (2000) between low and high cognitively demanding tasks, adapting this to enable disaggregation within our data sets where typically, learners were invited to interact within the teacher's demonstration of a procedure, but their contributing steps were restricted to mathematics that was already known to them. The preponderance of low demand did extend to application of known procedures, but rarely to higher cognitive demand as visible in terms of the extent of connections between and among concepts and procedures required. Movement towards generality and the extension of disciplinary thinking is also a function of what learners are asked and then able to do independently. Referring to the same lesson, the tasks for learners, notwithstanding the inclusive way in which the teacher asked learners for contributions, remained low levelled. When the product of binomials was introduced, and with it possibilities for learners to think about and then offer how they might extend the rule "multiply each term inside the bracket by the term outside the bracket", a procedure for this, based on the distributive law was demonstrated. As described elsewhere (Adler, 2017), extending learner disciplinary thinking through more cognitively demanding tasks within example sets remains a considerable challenge on the ground.

In the introduction to the MDI earlier in this chapter, we noted our attention to what we called "naming" and "legitimating" as the two instructional tools we use to analyse mediation through talk, or what we called "explanatory communication/talk". These draw on earlier work in which we theorized and described what was constituted *as* mathematics in pedagogic discourse (Adler & Davis, 2006; Davis, Adler & Parker, 2007). We drew on Bernstein's (2000, p. 36) insight that "key to pedagogic practice is continuous evaluation".[1] Whether explicit or not, talk in instruction will communicate messages (criteria in Bernstein's terms) as to what counts as mathematics. This would be both through the words used as well as to derivations of steps in procedures, or particular definitions, conventions and ways of proving, were substantiated (legitimated). While the purpose of our earlier work was to describe how notions about mathematics teaching and mathematics itself were communicated in mathematics teacher education, in MDI our purpose is to be able to describe whether and how mathematical principles are communicated to legitimate what counts as mathematics, and the extent to which learners are enabled to recognize and use the mathematics register in their talk.

Our motivation for this particular focus on talk is also a function of prevailing discourses in secondary mathematics teaching post-apartheid. At a more ideological level was the notion that mathematics must be relevant to everyday life, and language accessible to learners, ideological in the sense that these were seen as counter to an apartheid education which was alienating and inaccessible. What emerged in lessons were forms where everyday words or analogies used to substantiate mathematics that could be mathematically confusing, indeed, problematic (e.g. the distributive law a(b + c)= ab + ac was explained as "we share a equally with b and c"). At the level of the classroom practice was the widespread talk about mathematical objects and processes by teachers where reading out strings of symbols predominated, with little attention to either naming or elaborating the objects and operations being acted on – e.g. a(b + c) = ab + ac was read out as "a bracket b plus c equals a b plus a c" and thus did not draw attention to the operations in play. This obscures the description of the law – "the product of a and (b + c) is equal to the sum of the products ab and ac", or "we distribute multiplication over addition".

Drawing on Adler (2001) and Sfard (2008) and their work on language and mathematical discourse respectively, our analytic codes in MDI distinguish colloquial/informal from formal naming, with reading strings of symbols as between these. Legitimation codes distinguish firstly between non-mathematical and mathematical substantiations, with the latter further distinguished between those with partial or localized substantiations and those with full generality. The significance of these varying substantiations or criteria is the opportunity they open and/or close for learning. Most obvious are the extremes of legitimations based on the one hand on principles of mathematics, and thus with varying degrees of generality, and on possibilities for learners to reproduce or reformulate what they have learned in similar and different settings. On the other hand, appeals to the authority of the teacher, and so legitimations based on position (do this because I say so) and/or visual cues, produce a dependency on the teacher, on recall (this is what you must do); or on how things "look", requiring imitation that is local or situational and typically non-mathematical. While imitation might be necessary in aspects of mathematics learning (Vygotsky, *op. cit.*; Sfard, ibid), these cannot be the endpoint of learning. The criteria for what counts as mathematics that emerge over time in a lesson are thus key to what is made available to learn in terms of movement towards scientific concepts. We do not revisit the coding here, as details have been referred earlier. However, we need to point out that evaluation of the mathematical quality over a lesson assumes back-and-forth moment between the informal and formal, and partial and full general substantiations as inevitable in a lesson (in mediation), but that quality is indicated in overall movement in a lesson towards substantiations that are informed by mathematical principles, and the use of the mathematics register in talk.

In Figure 2.3 we present the evaluative part of the MDI framework that shows how we reach a summative judgement of a lesson in terms of each of the four elements here. The examples and legitimations indicate our interpretation of these and how they accumulate towards generality. The increasing task demands are related to more complex mathematical actions and naming towards more rigorous use of the mathematics register.

Object of learning = mathematical object and capability			
Examples	**Tasks**	**Naming**	**Legitimations**
The set of examples provides opportunities in the lesson for learners to experience:	Tasks provide opportunities for engaging the object of learning	Use of colloquial and mathematical words	Criteria for what counts as mathematics that emerge over time in a lesson and provide opportunity for learning geared towards scientific concepts.
E0 = NA No empirical cases	T0 = NA	N0 = NA	L0: all criteria are Non Mathematical (NM) i.e. Visual (V), Positional (P), or Everyday (E)
E1: one form of variation i.e. Similarity (S) or Contrast (C)	T1: carrying out known procedures only (K)	N1: NM – there is no focused math talk – all colloquial/everyday	L1: criteria include Local (L) – e.g. single case.
E2: at least two forms of variation: S and S OR S and C	T2: K and/or some application (A)	N2: movement between NM and Ms (words used as labels, talk about surface features), some Ma (mathematical language embedded in talks about objects) but mostly Ms	L2: criteria extend beyond NM and L to include Generality, but this is partial GP
E3: simultaneous variation (fusion - F)	T3: K and/or A and Use multiple concepts or make multiple connections (C/PS)	N3: movement between colloquial NM, Ms, but mostly formal math talk Ma	L3: GF math legitimation of a concept or procedure is principled and/or derived/proved i.e. full
	T2 → T1: A → K or C/PS → K is assigned to tasks set up at level 2 or 3 but then reduced to 1 when it unfolds.		

FIGURE 2.3   The evaluative part of the MDI framework.

Source: Adapted from Adler and Ronda (2015).

5. Theoretical/methodological overlaps and contrasts relating to generality

In this section, we focus on the overlaps and contrasts in the ways that the goal of generality is conceptualized across key mediating forms and in methodological approaches in the MPM and MDI frameworks.

Example spaces are coded for generality in the MDI through attention to the Variation Theory categories of similarity, contrast and fusion. In the MPM, while tasks and examples are noted, the example space itself is not coded for the extent of potential for generality, due to the evidence of limited purposive selection, and the evidence, too, of poor coherence and connection in the example space that sometimes negated the potential of the example space to mediate learning. This difference reflects contextual differences in the policy terrain, with recent policy directives in the primary mathematics terrain moving towards prescribing example spaces and feeding into our attention to mediation of these example spaces, and beyond these example spaces in instruction.

Problems with the coherence and connection of example spaces in some of the materials disseminated via policy in the secondary mathematics terrain have contributed to attention in the MDI model on example spaces themselves. Both approaches draw on the writing on example spaces and dimensions of possible variation, with different Variation Theory concepts drawn on for mediating example spaces for generality and sophistication/efficiency through a disciplinary lens: similarity, contrast and fusion in MDI and an adapted version of the range of permissible change in MPM.

Talk is attended to as an important mediational form in both frameworks. Language per se in terms of colloquial/formal discourse is attended to in MDI, together with the non-mathematics/mathematically principled substantiations or legitimations of what counts as mathematics. In MPM, given that primary level induction into mathematics involves emphasis on building connections between everyday and mathematical discourses, less emphasis is placed on *transitions* in word use from the colloquial to the formal. Instead, attention is focused on the extent to which discourse is mathematically coherent, connected and responsive, with all of these routes offering routes into generality. There is common attention across both frameworks to localized and more general instructional discourse, based on the semiotic bundle talk strands in MPM and to mathematically principled legitimation in MDI.

Methodologically, there are also important differences. While in the MPM, the levels in each strand represent trajectories towards generality, in the MDI, the aspects in focus are arranged in categories rather than in a hierarchy. Quality is then inferred on the basis of extent of generality seen in coherent movement between these categories. Further, while coding at the episode level is analytically possible in the MPM model (though evaluatively, it has been used only at the lesson level), in the MDI, inferences of quality are made only cumulatively at the lesson level. Across both frameworks currently, therefore, it is at the lesson level that changes in teaching for generality over time are considered.

6. Concluding comments

The MPM and MDI frameworks have provided insight into ways of considering trajectories from the localization seen on the ground, into the generality that is valued as a key aspect of mathematical working. Theoretically and developmentally, building this kind of trajectory is useful, with the developmental aspects grounded in our contextual realities. This means that the theoretical moves we have made are well attuned to the realities of instruction in low performance and poor resourcing contexts, though of course these moves may well be applicable elsewhere also. Additionally, as theoretical moves, they may well have broader purchase. Importantly in relation to the orientations that guide the dominant research paradigms in the mathematics education field, no demands are made in the MPM and MDI models for pedagogic forms related to learner-centredness. The attunement instead is to disciplinary trajectories that build towards generality, drawing from and adapting theories that centre on this idea. These trajectories guide our teacher development activities with primary and secondary teachers, and in the process we continue to learn about the possibilities and constraints in local contexts and conditions, for moving towards mathematics understood as "scientific" and the kinds of instructional skills required to bring this aspiration into being.

Note

1 Bernstein's notion of "evaluation" is not to be conflated with assessment.

References

Adler, J. (2001). *Teaching mathematics in multilingual classrooms*. Dordrecht: Kluwer Academic Publishers.

Adler, J. (2017). Mathematics teachers' "take-up" from professional development. In Th. Zachariadis, D. Potari, & G. Psycharis (Eds.), *Proceedings of the Seventh Greek Conference of the Greek Association of Researchers of Mathematics Education: Mathematical knowledge and teaching practices* (pp. 24–35). Athens: GARME. Retrieved from http://enedim7.gr/

Adler, J., & Davis, Z. (2006). Opening another black box: Researching mathematics for teaching in mathematics teacher education. *Journal for Research in Mathematics Education, 37*(4), 270–296.

Adler, J., & Ronda, E. (2015). A framework for describing mathematics discourse in instruction and interpreting differences in teaching. *African Journal of Research in Mathematics, Science and Technology Education*. doi:10.1080/10288457.2015.1089677

Adler, J., & Ronda, E. (2017). Mathematical discourse in instruction matters. In J. Adler & A. Sfard (Eds.), *Research for educational change: Transforming researchers' insights into improvement in mathematics teaching and learning* (pp. 64–81). Abingdon: Routledge.

Adler, J., & Venkat, H. (2014). Teachers' mathematical discourse in instruction: Focus on examples and explanations. In H. Venkat, M. Rollnick, J. Loughran, & M. Askew (Eds.), *Exploring mathematics and science teachers' knowledge: Windows into teacher thinking* (pp. 32–146). London: Routledge.

Alexander, R. (2000). *Culture and pedagogy: International comparisons in primary education.* Oxford: Blackwell.

Arzarello, F. (2006). Semiosis as a multimodal process. *Relime,* Numéro Especial, 267–299.

Askew, M. (2019). Mediating primary mathematics: Measuring the extent of teaching for connections and generality in the context of whole number arithmetic. *ZDM: The International Journal of Mathematics Education.* https://doi.org/10.1007/s11858-018-1010-9

Askew, M., Venkat, H., Abdulhamid, L., Mathews, C., Morrison, S., Ramdhany, V., & Tshesane, H. (2019). Teaching for structure and generality: Assessing changes in teachers mediating primary mathematics. In M. Graven, H. Venkat, A. Essien, & P. Vale (Eds.), *Proceedings of the 43rd Conference of the International Group for the Psychology of Mathematics Education* (Vol. 2, pp. 41–48). Pretoria, South Africa: PME.

Bernstein, B. (2000). *Pedagogy, symbolic control and identity: Theory, research and critique* (rev. ed.). Oxford: Rowman and Littlefield.

Cai, J., & Nie, B. (2007). Problem solving in Chinese mathematics education: Research and practice. *ZDM: The International Journal of Mathematics Education, 39*(5–6), 459–474.

Davis, Z., Adler, J., & Parker, D. (2007). Identification with images of the teacher and teaching in formalized in-service mathematics teacher education and the constitution of mathematics for teaching. *Journal of Education, 42,* 33–60.

Goldenberg, P., & Mason, J. (2008). Spreading light on and with example spaces. *Educational Studies in Mathematics, 69*(2), 183–194.

Gray, E. (2008). Compressing the counting process: Strength from the flexible interpretation of symbols. In I. Thompson (Ed.), *Teaching and learning early number* (pp. 82–94). Maidenhead, UK: Open University Press.

Gu, F., Huang, R., & Gu, L. (2017). Theory and development of teaching through variation in mathematics in China. In R. Huang & Y. Li (Eds.), *Teaching and learning mathematics through variation: Confucian heritage meets Western theories* (pp. 13–41). Rotterdam, The Netherlands: Sense Publishers.

Hoadley, U. (2007). The reproduction of social class inequalities through mathematics pedagogies in South African primary schools. *Journal of Curriculum Studies, 39*(6), 679–706.

Karpov, Y. V., & Bransford, J. D. (1995). L.S. Vygotsky and the doctrine of empirical and theoretical learning. *Educational Psychologist, 30*(2), 61–66.

Kieran, C. (2018). Seeking, using, and expressing structure in numbers and numerical operations: A fundamental path to developing early algebraic thinking. In C. Kieran (Ed.), *Teaching and learning algebraic thinking with 5- to 12-year-olds* (pp. 79–105). ICME-13 Monographs. Cham, Switzerland: Springer. https://doi.org/10.1007/978-3-319-68351-5_4

Kozulin, A. (2003). Psychological tools and mediated learning. In A. Kozulin, B. Gindis, V. S. Ageyev, & S. S. Miller (Eds.), *Vygotsky's educational theory in cultural context* (pp. 15–38). Cambridge: Cambridge University Press.

Leinhardt, G. (1990). *Towards understanding instructional explanations.* Washington: Office of Educational Research and Improvement. Retrieved November 1, 2016, from http://files.eric.ed.gov/fulltext/ED334150.pdf

Marton, F., & Booth, S. (1997). *Learning and awareness.* Mahwah, NJ: Lawrence Erlbaum Associates.

Marton, F., & Pang, M. F. (2006). On some necessary conditions of learning. *The Journal of the Learning Sciences, 15*(2), 193–220.

Marton, F., & Tsui, A. (2004). *Classroom discourse and the space of learning.* Mahwah, NJ: Lawrence Erlbaum Associates.

Mason, J., Graham, A., & Johnston-Wilder, S. (2005). *Developing thinking in algebra.* London: Paul Chapman Publishing.

Mathews, C., Venkat, H., & Askew, M. (2018). Primary teachers' semiotic praxis: Windows into the handling of division tasks. In N. Presmeg, L. Radford, W. M. Roth, & G. Kadunz (Eds.), *Signs of signification* (pp. 257–274). ICME-13 Monographs. Cham, Switzerland: Springer. https://doi.org/10.1007/978-3-319-70287-2_14

Reeves, C., & Muller, J. (2005). Picking up the pace: Variation in the structure and organization of learning school mathematics. *Journal of Education, 37,* 103–130.

Sfard, A. (2008). *Thinking as communicating.* Cambridge: Cambridge University Press.

Skovsmose, O. (2011). *An Invitation to critical mathematics education.* Rotterdam, The Netherlands: Sense Publishers.

Stein, M., Smith, M., Henningsen, M., & Silver, E. (2000). *Implementing standards-based mathematics instruction: A casebook for professional development.* New York: Teachers College Press.

Treffers, A. (2001). Grade 1 (and 2): Calculation up to 20. In M. van den Heuvel Panhuizen (Ed.), *Children learn mathematics* (pp. 43–60). Utrecht, The Netherlands: Freudenthal Institute.

Venkat, H., & Adler, J. (2012). Coherence and connections in teachers' mathematical discourses in instruction. *Pythagoras, 33*(3). http://dx.doi.org/10.4102/pythagoras.v33 i3.188

Venkat, H., & Adler, J. (2017). Frameworks supporting the coding and development of mathematics teachers' instructional talk in South Africa. In S. Zehetmeier, B. Rösken-Winter, D. Potari, & M. Ribeiro (Eds.), *Proceedings of the Third ERME Topic Conference on Mathematics Teaching, Resources and Teacher Professional Development* (pp. 228–237). ETC3, October 5–7, 2016. Berlin, Germany: Humboldt-Universität zu Berlin.

Venkat, H., & Askew, M. (2017, February 1–5). Focusing on the "middle ground" of example spaces in primary mathematics teaching development in South Africa. In T. Dooley & G. Gueudet (Eds.), *Proceedings of the Tenth Congress of the European Society for Research in Mathematics Education* (pp. 3193–3200). CERME10. Dublin, Ireland: DCU Institute of Education and ERME.

Venkat, H., & Askew, M. (2018). Mediating primary mathematics: Theory, concepts and a framework for studying practice. *Educational Studies in Mathematics, 97*(1), 71–92.

Venkat, H., & Naidoo, D. (2012). Analyzing coherence for conceptual learning in a Grade 2 numeracy lesson. *Education as Change, 16*(1), 21–33.

Venkatakrishnan, H. (2005). The implementation of the mathematics strand of the key stage 3 strategy: A comparative case study: Summary of findings. *Research Intelligence, 93,* 20–22.

Vygotsky, L. (1987). Thinking and speech. In N. Minick, R. W. Rieber, & A. S. Carton (Eds. & Trans.), *The collected works of L.S. Vygotsky, Volume 1: Problems of general psychology.* New York: Plenum.

Watson, A., & Mason, J. (2006). Seeing an exercise as a single mathematical object: Using variation to structure sense-making. *Mathematical Thinking and Learning, 8*(2), 91–111.

Zodik, I., & Zaslavsky, O. (2008). Characteristics of teachers' choice of examples in and for the mathematics classroom. *Educational Studies in Mathematics, 69*(2), 165–182.

3

THE ROLE OF TEACHERS' KNOWLEDGE IN THE USE OF LEARNING OPPORTUNITIES TRIGGERED BY MATHEMATICAL CONNECTIONS

Genaro de Gamboa, Edelmira Badillo, Miguel Ribeiro, Miguel Montes & Gloria Sánchez-Matamoros

1. Introduction

The making of connections is a linchpin of mathematics education, as it is related to the development of a broader and deeper knowledge of mathematics (Skemp, 1971; Triantafillou & Potari, 2010). Mathematics, and in particular mathematics problem-solving, is characterized by the interconnectivity between different content areas (e.g. algebra and geometry), between different representations or different procedures and between mathematics and outside-mathematics situations. Therefore, connections play a major role at all educational levels, especially in primary school, where traditionally formal mathematical education begins.

Following a research trend in recent years, several perspectives on connections have been developed, both regarding the ways connections occur in mathematics (e.g. Zazkis & Mamolo, 2011) and the way connections are established in the classroom (e.g. De Gamboa & Figueiras, 2014; Montes, Ribeiro, Carrillo & Kilpatrick, 2016). We focus on the latter perspective and on the categorizations that emerge which describe connections as a complex system of relationships in which outside-mathematics situations, systems of representation and/or heuristics are linked.

Making connections in the classroom can help students to identify new applications of mathematics to real-world problems and to give meaning to such problems in school contexts. It may also make it easier to use different representations when solving problems, as well as to reinterpret and rebuild connections between concepts, properties and/or procedures. Consequently, promoting the emergence of connections in the classroom has the potential to trigger a wide array of learning opportunities. The usefulness of these opportunities depends on teachers' knowledge and on their ability to identify, interpret and promote learning opportunities stemming from connections, and to make decisions during the classroom activity that help students to build up a broader and deeper mathematical knowledge.

The presence of connections when conceptualizing teachers' knowledge – assuming a practice-based perspective – reveals a relationship between teachers' knowledge and the way connections are established and used in the classroom. For instance, Rowland, Turner, Thwaites and Huckstep (2009) consider connections as a domain of teachers' knowledge that refers to teachers' ability to anticipate complexity, make decisions about sequencing, make connections between procedures and make connections between concepts. As regards Ball, Thames and Phelps (2008), connections are related to teachers' awareness of how mathematical concepts are connected throughout school years.

In managing the development of students' understanding, teachers need to mobilize their knowledge (both mathematical knowledge and pedagogical content knowledge) in a very specialized way. In order to capture the nuances of that knowledge and to characterize the specialized features of mathematics teachers' knowledge in terms of what is mobilized when promoting and exploring mathematics learning opportunities arising from mathematical connections, we consider such specialization in the sense of the framework of the Mathematics Teachers' Specialized Knowledge (MTSK; Carrillo et al., 2018).

With a view to deepening our understanding of connections and their role in practice (potentialities and constraints), and to conceptualize ways of improving the effectiveness and utility of such connections in terms of teacher knowledge, the concretization of the aforementioned categorizations of connections need to be studied and expanded in relation to several mathematical topics. Amongst the diversity of topics in school mathematics, measurement is perceived as a crucial element in pupils' development of mathematical understanding and knowledge (e.g. Sarama, Clements, Barret, Van Dine & McDonel, 2011). It is also a rich environment in relation to the emergence of mathematical connections, as it represents a natural linkage between numbers and operations, geometry and real-world problems. In particular, the introduction of the measurement of length – its different dimensions (Clements & Sarama, 2007) and properties – can enact several aspects of mathematical connections related to natural and rational numbers, different representations of numbers, the procedures related to the measurement of length and different units of measurement (e.g. Szilágyi, Clements & Sarama, 2013).

With the aim of gaining a better understanding of what features of teachers' knowledge can help teachers to foster the making of connections in the classroom and to make the most of these connections in terms of exploiting the learning opportunities stemming from them, in this chapter we present the case of Carla, a prospective teacher developing an introduction to standard length units in the second grade of primary school. We start by characterizing mathematical connections that emerge during Carla's lesson, and we identify and discuss the learning opportunities stemming from those connections. Then, we analyse what features of her knowledge are related to the use of connections, and what other features of teacher knowledge would have helped her to effectively use connections to build a deeper and broader knowledge of length measurement.

2. Length measurement teaching and learning

The main ultimate aim in performing a measurement is to assign a numerical value to an object's attribute. Before measuring, we need to identify which of the object's properties are measurable and differentiate them from those that are not (Campbell, 1928). When assigning numerical values to measurable properties, we establish a relation between the property that is being measured and the mathematical properties of the numbers that are used for measuring. Measurement fosters the establishment of relationships between geometrical concepts and numbers, by analysing shape, position, symmetries, rotations and translations using the numerical properties of measurable magnitudes such as angles, distances, lengths and areas. This relation between geometrical properties and numerical properties also allows us to explore the relationship between measurable magnitudes such as perimeter and area when shape or position is changed.

The richness of measurement in terms of the relations between geometry and numbers is related to a high level of complexity. With a view to gaining a better understanding of how connections can be established when working with length measurement in the classroom, an exhaustive analysis of the mathematical elements involved can be useful. As in many mathematical topics (e.g. adding or dividing fractions), understanding the steps involved may be more complex than the process itself, as it involves several stages including many key concepts that are articulated through the connections between problems, representations, definitions, properties and procedures.

Stephan and Clements (2003) posited that six key concepts must be mastered to develop a full understanding of measurement and the skills required for it: (1) equal partitioning – the mental process of dividing an object into equal parts, requiring the acknowledgement of the divisibility of the object; (2) unit iteration – the skill of exhaustively repeating the unit successively to cover the object; (3) transitivity – recognition of the mathematical property of measure that ensures that if the measure of A is bigger than the measure of B and the measure of B is bigger than the measure of C, then the measure of A is bigger than the measure of C; (4) conservation of the measure through rigid transformations that do not change the amount of magnitude; (5) addition and accumulation of distance – recognition that the measurement process outcome is the measure of the object (how many units have to be repeated to equal the measurement of the object) and (6) relationship between number and measure, implying acceptance that a variation of the unit of measure would generate a change in the measurement outcome (total amount of units).

Length measurement can be divided into several mathematical packs that include conceptual and procedural elements, as proposed by Ma (1999). In the context of classroom activity, the comprehension of length measurement is associated with that of a system of practices that allow problems related to measure to be solved (Rondero & Font, 2015). In the case of length measurement in primary school, we use an epistemic configuration of length measurement to analyse its complexity (Rondero & Font, 2015) and how the six previous concepts are supposed to be mastered by pupils. Specifically, we apply the six levels of complexity proposed by Rondero and Font (2015) to length measurement in the early years of primary school which become useful to structure the discussion around the teaching and learning of measurement, as will be shown later in this chapter.

The first level refers to problematic situations related to the need to establish a universal method for the measurement of length in a quotidian context. The second level of complexity concerns the different representations associated with the measurement of length. The third level of complexity consists of the definitions and concepts related to the measurement of length. The fourth level refers to propositions and properties of the concepts related to the measurement of length. The fifth level is formed by the procedures used when performing a length measurement. Finally, the sixth level is related to the arguments that can be used when interpreting the results of length measurements and the decision-making stemming from those results.

The specificities of this framework serve to interpret teachers' and students' interventions in terms of the construction of knowledge of length measurement more explicitly than through the six key concepts proposed by Clements and Sarama (2007). These six levels are used when discussing the nature of the connections employed by teachers and those emerging from the answers and/or comments students make in response to different problems of length measurement. Moreover, the different elements in each level allow us to identify learning opportunities stemming from connections in the classroom.

In the analysis we focus on the mathematical complexity of the measurement of length in early stages and its relationships with the different subdomains of the MTSK conceptualization.

3. Connections

The mathematical connections that are present in the classroom activity are bonds between different mathematical ideas. In particular, we can define mathematical connections as a network of links that coordinate definitions, properties, procedures and/or representations by means of coherent and logical relations (De Gamboa & Figueiras, 2014). The establishment of such connections implies, in most cases, the construction of complex structures between the different links (De Gamboa, 2015). In this way, coordination between the links that form a connection entails the assessment of previous connections, or the creation of new ones.

In a classroom context, mathematical connections occur – among other classroom situations – when students and teacher interact, generating linear and

TABLE 3.1 Mathematical complexity of the measurement of length in early stages.

Problematic situations

PS1: How to compare lengths indirectly?

PS2: How to establish numerical comparisons between lengths?

PS3: How to communicate quantities of length in a universal way?

PS4: What is the most exact way of communicating lengths?

PS5: What is the best way of subdividing a unit of length?

PS6: How to establish equivalencies between units of measurement?

PS7: How to compare measures made with different units of measurement?

Definitions and concepts

D1: Cardinal and ordinal numbers.

D2: Natural and positive rational numbers.

D3: Approximation and estimation.

D4: Occupied space and comprised space.

D5: Distance.

D6: Units and subunits of measurement.

D7: Standard and non-standard units of measurement.

D8: Perimeter of plane shapes.

Procedures

PR1: Addition of positive quantities.

PR2: Multiplication of positive quantities.

PR3: Representation of quantities on the number line.

PR4: Estimation of lengths.

PR5: Strategies of mental calculation.

PR6: Comparison of lengths.

PR7: Discretization of rational quantities.

PR8: Use of instruments of measurement.

PR9: Definition of an algorithm of the process of measurement (choose a unit of measurement, extend the unit repeatedly and exhaustively along the object, count the number of iterations needed to cover the length of the object, approximate the results at a certain level of exactness, assess the plausibility of the result obtained).

Representations

R1: Verbal representation using adjectives like "long" or "short" and adverbs like "more" or "exactly".

R2: Symbolic representations using natural, rational and irrational numbers.

R3: Graphic representations on the number line.

R4: Enactive representations using parts of the body.

Properties and propositions

P1: Transitivity of measure.

P2: Transitivity of numbers.

P3: Conservation of length.

P4: Accumulation and additivity: recognition that the outcome of the measurement process is the measure of the object.

P5: Inverse relation between unit's length and the numerical outcome.

P6: Equivalence between procedures if results differ by less than a given percentage.

Arguments

A1: When a length of measurement and its fractions are defined, numerical comparisons can be established.

A2: Units of length have to be well known (e.g. hand span, arm or foot) so the communication of length measurement can be effective.

A3: The most exact way to communicate lengths is through the use of consistent units (so the length of the unit does not depend on the particular instrument used).

A4: The suitability of the unit's partition depends on the context in which the unit is used. The more exactness is needed, the smaller the subunits have to be.

A5: In order to establish equivalencies between units, they should be consistent. The exactness of such equivalencies depends on the possibility of measuring 1 unit using the other one and obtaining a natural, rational or irrational result.

A6: There is an inverse relationship between the length of the unit and the numerical outcome.

non-linear chains of links, considering the varying nature of the interactions that happen in a classroom. The complexity associated with the construction of the connections can sometimes generate misinterpretations or incompleteness of the links produced, which is related to common mistakes in school mathematics (De Gamboa, 2015). Therefore, the emergence of connections in a class generates learning opportunities related to the possibility of reorganizing conceptual structures and reorienting the misinterpretations based on establishing new links and restructuring the existing ones.

When thinking about the kinds of connections, two major dimensions can be perceived: intra-mathematical and extra-mathematical. Intra-mathematical connections are produced in a mathematical context where only mathematical representations, properties, procedures and arguments are used. Extra-mathematical connections, on the other hand, are connections that link mathematical concepts – from a broader perspective, including definitions, properties, procedures and representations – and problematic situations in a non-mathematical context.

Intra-mathematical connections can be (a) *conceptual connections* (implying treatment or implying conversions) or (b) *transversal-process* connections. Conceptual connections consist of the relationships that are established between representations, procedures or techniques associated with a single concept or different concepts, while transversal-process connections are the relationships between a mathematical concept and a mathematical process that is transversal to different mathematical concepts. Specifically, we consider the connections with arguments, proof and heuristics in problem-solving.

The connections related to transversal processes are associated with the transition between a mathematical activity focused on algorithmic activities and a mathematical activity characterized by a deeper degree of abstraction, based on the identification of patterns, the justification and proof of results and the rigorous communication of mathematical information. For example, while observing during a classroom activity that there is an inverse relationship between the measurement unit used and the numerical outcome in the measurement process, a connection related to processes would establish a relation between the concrete relationship identified and the process of justification of that property, with emphasis on its generalization.

Conceptual connections can be divided into two kinds, (1) connections that imply treatment and (2) connections that imply conversions, in the sense of Duval (2006). Concerning (1), no changes of register occur, hence connections of this kind are the most common in the mathematics classroom because they are the connections associated with algorithmic procedures, such as the equivalence between measurement units. In the second case, changes of register do occur, implying the coordination of meanings between different registers of representation, such as the calculation of a perimeter starting from the picture of the geometrical shape, and the representation of this calculation with symbols and letters. This coordination between different registers of representation of the same context is one of the main elements for the understanding of a concept, because the transit between them allows us to distinguish the elements that are mathematically relevant from those that are not (Duval, 2006).

Extra-mathematical connections are characterized by connecting mathematical concepts with situations that: (a) have clearly different objectives from those of the mathematical activity in the classroom, (b) use a different kind of discourse from that used in mathematics classrooms and (c) require a set of symbols and a language that clearly differ from those used in mathematics (Walkerdine, 1988). Therefore, when extra-mathematical connections are established, the validity and coherence of the links are determined not only by mathematical rules, which imply a complex coordination between the usual concepts and procedures of validation used in mathematics, but also by the procedures of validation that are used in other disciplines.

In the mathematics classroom, connections happen mainly in two situations. Firstly, when the teacher plans the sequence of activities with the aim of making explicit connections, and secondly, in contingency situations in which a student's comment or the classroom discussion triggers the establishment of connections. In both cases, the ability of the teacher to manage and use the learning opportunities that the connections can provide is crucial. In order to explore and gain a better understanding of how connections can be used in the classroom as a learning opportunity, it is important to analyse teachers' knowledge.

4. Teachers' knowledge

Teachers' knowledge has been a central concern of mathematics education research over the past 30 years (e.g. Shulman, 1986). In particular, teachers' knowledge of measurement has been a focus of research (Olivero & Robutti, 2001; Lanciano, 2003; Steele, 2013), with the aim of exploring and characterizing it, as well as analysing the challenges related to measurement teaching and learning. Some of this research has addressed, in particular, prospective teachers' knowledge to teach measurement (Policastro, Almeida & Ribeiro, 2017; Subramanian, 2014). In this chapter we use the MTSK model (Carrillo, Climent, Contreras & Muñoz-Catalán, 2013; Carrillo et al., 2018), which proposes a specialized perspective on teachers' knowledge, to help us understand teachers' knowledge, here in particular in the scope of measurement.

This model follows the seminal reflection by Shulman (1986) that considers three domains – mathematical knowledge, pedagogical content knowledge and the affective domain (although this last domain will not be under scrutiny in this chapter) – as elements that, while possibly used by the teacher in a complex and integrated way, can be analysed separately in order to gain a deeper understanding of professional knowledge.

Concerning teachers' knowledge of measurement, with the focus only on mathematical knowledge, this model identifies three subdomains. Knowledge of Topics (KoT), in the context of measurement, would correspond to Knowledge of Measurement (KoM). This includes teachers' knowledge of the different mathematical elements that make up the topic of measurement: definitions, properties, procedures, registers of representations, phenomenology and applications. Here we can find research into, for example, teachers' knowledge of measurement of perimeter and area, as well as the difference between both (Steele, 2013). In this subdomain we can also find teachers' knowledge of how the different properties and representations of the measurement are related. Knowledge of the Structure of Mathematics (KSM) refers to the knowledge of the kind of connections in and related to the topic under discussion with other(s), or with the same topic with different levels of complexity. In the case of measurement, it comprises the knowledge of the different relationships of measurement with other mathematical topics, such as arithmetic (Meissner, 2011), numbers (Rafiepour & Karimianzade, 2017) and geometry (Fonseca & Cunha, 2011), or the relationship between the content in each topic worked on in the classroom and their connection to other topics at a higher mathematical level. Knowledge of the Practice of Mathematics (KPM) covers the knowledge of the processes of justification, proof and refutation that can appear when dealing with measurement, together with heuristic strategies of problem-solving that can be used concerning measurement. For instance, the relationship of the idea of proof with the precision of the measurement is of huge importance (Mariotti, 2011).

As regards pedagogical content knowledge, in the particular case of measurement, we can find three subdomains. Knowledge of the Mathematics Teaching (KMT) includes knowledge of tasks, to manage the learning of measurement, and methodological resources, whether physical, such as puzzles (Sensevy, 2009), or technological (Kortenkamp & Rolka, 2009). Here we can also find theories of teaching, both formal and personal, that help teachers to make sense of the approach adopted. One other subdomain of teachers' PCK concerns the Knowledge of the Features of Learning Mathematics (KFLM), encompassing teachers' knowledge of, among others, the elements into which we can unpack measurement learning, the usual ways in which students interact with measurement, usual mistakes and obstacles in the learning of the concept that occur and theories of learning measurement. Knowledge of Measurement Learning Standards (KMLS) is the knowledge of the degree of competence and performance that students are expected to reach in each grade, as indicated in curricula, or by professional associations (e.g. NCTM).

5. Methods and context

Data was collected as part of the 4th year of the initial teachers' training programme at the *Universitat Autònoma de Barcelona*. During the previous 3 years, prospective teachers had taken mathematics and mathematics education. In particular, one of those courses addressed teaching and learning magnitudes and measurement, dealing specifically with teaching and learning length and its measurement, focusing on the importance of making connections in the classroom. In the 4th year, prospective teachers have an internship of 240 hours in a primary school. Prospective teachers are assigned to a particular classroom where they design, implement and assess a teaching unit on a mathematical topic. One of the activities they are required to develop during the internship – as part of a course at university – is the analysis of a video-episode they select from their own practice. Prospective teachers are required to record some of their teaching moments in classes (a total of 10 hours) during the field practice. Then they have to choose one of these classes and identify what they consider to be a significant episode in terms of the mathematical content approach and students' understanding. To select the episodes, they should use Sherin, Linsenmeier and van Es's (2009) criteria: window, clarity and depth. These criteria have been discussed previously as part of a course aimed at discussing the use of such criteria in order to obtain powerful information about and for improving teachers' practices.

This interpretative study is framed in the qualitative research paradigm. Considering the nature of our research question, we use a case study (Bryman, 2004) to analyse connections, learning opportunities and teachers' knowledge in a context that has not been designed ad hoc for the purposes of this research. Amongst the several case studies that have been developed as part of a broader research project – aimed at identifying and deepening the understanding of the content of teachers' knowledge and how it intertwines with teachers' actions and beliefs when they choose and analyse videos from their own practice – five prospective teachers focused their intervention on the topic of length measurement. We focus on one of those prospective teachers, Carla, who was selected according to two criteria: (a) she made significant use of Sherin et al.'s (2009) criteria, which allowed her to conduct a detailed and profound analysis of her own practice, and (b) her design of the teaching unit was focused on the making of connections between several aspects of length measurement.

We have selected episode fragments from Carla's recorded sessions, with 7- and 8-year-old pupils, in which the three dimensions have been identified at a high level. Window dimension is related to explicit evidences of students' different levels of comprehension of length measurement. Depth dimension concerns interactions in which students take part in the decision-making about the units, the instruments and the procedures that can be used. Finally, evidences of the clarity dimension have been identified when students transparently express mathematical arguments about their comprehension of length measurement. The richness of the evidences obtained in the light of the three previous dimensions, along with the analysis of the mathematical complexity of length measurement in the context of non-standard units, are an ideal source of data for the analysis of how teachers' knowledge determines the use of connections in the classroom.

We focus on an episode involving three tasks exploring the use of non-standard units for measuring length in a "real life" (classroom) context – foot length, hand span and width of a finger to measure classroom length, the height of a book cover and the height of a glue pot respectively – as it helps to reveal Carla's knowledge of connections. The episode is first analysed with a focus on identifying mathematical connections. Afterwards we analysed the links that form the connections using the epistemic configuration (presented in Table 3.1). When analysing Carla's work, focusing on connections, we adopt the categorization proposed by De Gamboa and Figueiras (2014) to identify how emerging intra-mathematical connections sustain the extra-mathematical connection, in terms of their nature and their relationship with the task performed.

The characterization of connections using the epistemic configuration of length measurement allows us to interpret such connections in light of specific problems, representations, definitions, properties, procedures and arguments, which provides a framework for identifying learning opportunities in the context of length measurement. It is important to note that, although the analysis tends to focus on identifying missed instruction opportunities, such situations are perceived as learning opportunities, and as a powerful tool for use in education. Thus, their inclusion enhances the discussion presented in the subsequent sections.

Finally, we analyse how teachers' knowledge can help/constrain students to make the most of the learning opportunities that emerge within the classroom. To analyse teachers' knowledge, we use the MTSK conceptualization which serves to focus, in depth, on the measurement-related knowledge Carla employed. This model allows us to identify concrete aspects of Carla's knowledge used in the episode, enhancing our understanding of the specialized knowledge supporting the management of connections as well as some specificities of her knowledge that could have helped her to use some missed learning opportunities. The episode we analyse here concerns revisiting three activities about the use of non-standard measures. Analysing the episode enables us to identify three blocks of connections, as we show in the following section.

6. Analysis and discussion

The three different tasks involved measuring lengths in a "real life" context, establishing an extra-mathematical connection related to a problem that needs to be tackled from a mathematical perspective (how can we compare and represent lengths?). This connection relates length to the numerical value representing it, and it is based on comparing the particular length to be measured with the length of an object that can be used as a reference (hands, feet, etc.), referred to as the unit of measurement.

The analysis showed that eight other particular connections were established in relation to the general extra-mathematical connection. These connections can be grouped in three different blocks defined by the elements of the mathematical complexity of length measurement that is highlighted. Firstly, a block of three connections involving the inverse relation between the length of the instrument and the numerical outcome was identified. Secondly, three connections related to alternative procedures for length measurement form another block. Finally, two connections between standard and non-standard units form the third block. The details of each of the connections in each block regarding the mathematical complexity of measurement and the teachers' knowledge related to the use of the learning opportunities triggered by connections are discussed here.

In the first block, an initial connection is identified when reviewing the results of the measurement of the height of a book cover using hand span. The results obtained by the students are similar but not the same (different hand spans). The connection occurs when the teacher asks "What happened when we measured?" to which a student answers that they are all wrong, in response to whom the teacher remarks that the results are not wrong. A link between the set of results and their mathematical validity is established, based on the need to have a criterion to decide whether two different results of the same measurement are equivalent or not. This link is related to problematic situations PS2, PS3 and PS4 which refer to comparison, communication and exactness in the measurement of length.

The difference between the results (the numerical answer) reported by the students can be tied to two characteristics of measurement. On the one hand, the approximate character of measurement, along with the possibility of expressing the numerical result using natural or rational numbers, or even imprecise expressions such as "a little bit more", means a margin of error needs to be created to consider two results as equivalent. On the other hand, the difference in the results may be due to the inconsistency of non-standard units, with different results obtained depending on the instrument being used.

When the teacher asks the students why the results are different, two students (Daniel and Hugo) answer by formulating a hypothesis regarding the inverse relation between the length of the instrument and the numerical outcome. Therefore, a link is made between the validity of the results and the formulation of a hypothesis regarding the dependence of the result on the length of the instrument. The formulation of this hypothesis is related to the arguments referring to the exactness of measurement – A2: units of length have to be well known so the communication of length measurement can be effective; and A3: the most exact way of communicating lengths is through the use of consistent units (so the length of the unit does not depend on the particular instrument used).

Finally, the teacher justifies the validity of the hypothesis by an enactive verification (R4: enactive representations using parts of the body) comparing her own hand span with a student's hand.

TEACHER: Can anyone tell me why we have obtained different results? Let's see, Daniel . . .

DANIEL: Because one of us has a big hand, a small hand . . .

HUGO: Each person has a different sized hand, and the bigger the hand, the fewer hands fit. And the smaller the hands, the more hands fit.

The inverse relationship between the length of the instrument and the numerical outcome (property P5) appears again when reviewing the results of measuring the length of the class using their feet and the height of a pot of glue using the width of their fingers. In both cases a second and a third connection are identified when

Daniel emphasizes the relationship between the length of the feet and the numerical outcome, and the relationship between the width of the fingers and the numerical outcome. In both cases the teacher makes enactive verifications, showing the difference between the space occupied by her feet and fingers in comparison with those of the students. Finally, at the end of the third connection, the teacher formulates a question related to the reliability of the measures being reviewed, emphasizing the connection between the inverse relationship (P5) and the initial question related to the validity of a set of different results when measuring the same object using the same unit of measurement. However, the latter intervention goes beyond the validity of the results, drawing attention to the lack of reliability of the results obtained when using non-standard units, and therefore to the usefulness of standard units.

The analysis of these three connections shows how elements from several levels of the complexity of measurement at early stages (Table 3.1) appear in the classroom activity. The teacher's management of classroom activities may foster the use of learning opportunities stemming from the previous three connections. The content of KPM can contribute towards helping the teacher to make remarks regarding the validation and verification of results and the formulation of hypotheses, as the mathematical rules to justify whether or not two results are the same or equivalent in a measurement context are different from those used in arithmetic. In this sense the enactive verification of this relation carried out by the teacher should lead to more elaborate justifications. KSM serves to make recursive remarks in the approximate nature of measurement since the use of decimal expressions and fractions may be needed in order to represent real length measurements. In relation to the problems generated by the inconsistency of non-standard units, the teachers' KoT – in the particular scope of measurement – along with Knowledge of the Features of Learning Measurement can help the teacher to address usual mistakes and obstacles related to the use of nonstandard units by emphasizing that the numerical result of the measurement process depends on the instrument used.

The three connections are therefore intra-mathematical connections with conversion, as there are changes from a numerical and symbolic register to an enactive and physical register, formed by different links. The fact that all three connections rely on a change of register of representation underlines the importance of the teacher knowing different registers of representation (KoT), how the physical and numerical representations are related (KSM), what resources she can use to make this correspondence of registers clear (KMT) and how this relation is also important in other school topics such as area and volume (KMLS).

During the review of the measurement of the height of the book cover using the hand span and the height of a pot of glue using fingers, three connections related to procedures were identified. In the case of the height of the book cover, Isaac asks if it is also correct to do the measurement by opening the hand partially.

ISAAC: Can we do it like this as well? [Opening his hands completely.]
TEACHER: Of course, another thing is how we place our hands. If some of you place them like this [partially opened] and some of you place them as Isaac

has [completely open], Isaac will get less. . . . But it doesn't mean that it is wrong; it simply indicates that we have different hands and we have measured differently.

Isaac's intervention and the teacher's answer establish a fourth connection between the correct procedure for the measurement of length and an alternative procedure. The teacher's explanation remains only at a superficial level, thus not allowing students to deepen their knowledge (and conception) of length. In fact, by saying "it doesn't mean that it is wrong", she is likely confusing the students, who consequently may not appreciate the importance of following the same procedure when measuring length. The main relation at the core of this intra-mathematical connection with treatment is between the standard procedure and an alternative one. While the standard procedure (completely extended hand) can be replicable if it is performed by the same person, in the case of the alternative procedure, replicability is much more difficult, and therefore they cannot be equivalent procedures.

This connection is related to procedure PR9 in the epistemic configuration of length measurement, concretely, to the step "extend the unit repeatedly and exhaustively along the object" that is being measured. Therefore, the connection triggers an opportunity for discussing the importance of using well-defined procedures in measurement in order to establish reliable comparisons between the results of measurements, related to the problematic situation on how to establish numerical comparisons between lengths (PS2).

The teacher's confusion is a lost opportunity for clarifying two possible reasons for different results. In the first block of connections, the difference was due to the inverse relation between the length of the instrument and the numerical outcome, while in this case the different results are generated by a lack of accuracy when performing the procedure of measurement. The teacher's incorrect answer can create a false link between the inverse relationship discussed in the first block of connections and the lack of accuracy in the performance of the procedure.

During the discussion of the results of measuring the length of the classroom using their feet, another student (Miguel) asks if separating their feet while performing the measurement would produce a different result.

MIGUEL: If we separate our feet we are not doing [measuring] anything.
TEACHER: Of course, we have to put them next to each other.

Moreover, Miguel does an enactive representation of his question, not only separating his feet while measuring but also changing the direction of his feet. Thus, another intra-mathematical connection with treatment between the standard procedure and an alternative one is established. However, the teacher fails to take this opportunity to continue exploring the sub-connection by emphasizing that the procedure must be performed exhaustively (intuitive introduction of the notion of algorithm), in order to obtain the same answer when using the same measurement unit and representational system.

The third intra-mathematical conceptual connection with treatment between procedures is detected during the discussion of the way students measured the height of a pot of glue using their fingers (Ribeiro, Badillo, Sánchez-Matamoros & Artès, 2016). While most students obtained numbers close to 10, Miguel's answer was one. When Carla asked him to explain how he obtained that result, Miguel elucidated his reasoning by placing his finger perpendicularly to the base of the pot, while the rest of the class had placed their fingers parallel to the base:

TEACHER: No? How many fingers did you get?

MIGUEL [placing his finger vertically next to the glue pot]: One!

TEACHER: One? Like this? [The teacher repeats the measurement process using the index finger horizontally] . . .

MIGUEL: Ah, four, four . . .

TEACHER: No! It can't be . . . you should get eight; you are doing it wrong.

This situation provides an excellent opportunity to discuss the importance of establishing a common procedure when conducting measurements. In this case, Miguel's answers show the use of non-standard units in a non-standard way. However, Carla's arguments and exemplification indicate her sole focus on the standard measurement process. In this sense, she seems to be taking for granted the underlying procedures, thus disregarding an answer that differs from her own (Jakobsen, Ribeiro & Mellone, 2014). Finally, she also fails to grasp (or at least discuss with her students) the relationship between the number obtained and the measurement method used.

These three latter connections are closely related to the sequence of steps that have to be followed to perform a measurement correctly (PR9) from the epistemic configuration of length measurement. They are also related to the "universal" communication of lengths and property (P3) as well as to accumulation and additivity in the measurement of length (PR4). The opportunities stemming from the latter three connections can be addressed by, on the one hand, conducting discussions in the classroom about the importance of unifying procedural criteria and using students' interventions to emphasize important elements in the comprehension of measurement using a deep knowledge of the procedures involved in length measurement and their characteristics (KoT). Although in the first three connections the teacher shows a sound knowledge of the inverse relation between the length of the instrument and the numerical outcome, in the fourth connection the teacher shows a lack of knowledge of the importance of following the steps in the measurement procedure thoroughly (PR9), which could have caused confusion in the students, reinforcing common misconceptions in primary school (Sisman & Aksu, 2016). On the other hand, the identification and interpretation of common mistakes in students' answers and comments based on knowledge of common mistakes and obstacles in learning length measurement (KFLM) may help the teacher to find the suitable moments to emphasize key concepts in length measurement. Knowledge of the Features of the Learning Mathematics, in the particular case of Measurement (KFLM), also appears to be linked to knowing, identifying and anticipating students' difficulties and misconceptions. Making use of students' interventions can be a good opportunity to emphasize mathematical contents that can be problematic to them.

Finally, two connections between the concepts of standard and non-standard units are identified. The first is established during the review of the measurement of the height of a book cover, when one of the students describes his measurement as two hand spans and one centimetre. This conceptual intra-mathematical connection with treatment between the concepts of hand span and centimetre is related to the problematic situations related to the communication of lengths, its exactness, the ways of subdividing units of measurement and the establishment of equivalencies between units (P3, P4, P5 and P6 respectively). Moreover, the establishment of this connection is related to the numerical and enactive representations of length, the definitions of standard and non-standard units, the property of accumulation and additivity and the standard procedure of measurement.

The relationship between the use of standard and non-standard units appears again when reviewing the measured height of the pot of glue, but from a broader perspective. At the end of the review the teacher reflects upon the usefulness of the non-standard units.

TEACHER: Did I really ask you to measure with your feet, hands and fingers? Would we obtain the same results? . . . We need to find another way, another system so we can measure and obtain the same results, don't we?

An intra-mathematical connection with treatment between standard and non-standard units is established. This connection is related to the problem of using non-standard units that started the first connection in the first block, and it sums up the implications of using non-consistent units, since the inverse relation between the length of the instrument and the numerical outcome implies that the results may differ considerably if different people perform the measurement. Therefore, this connection draws attention to the extra-mathematical connection that encompasses the whole episode and the other seven connections identified.

This connection triggers an opportunity to show the importance of establishing consistent units that can be divided into subunits, and of defining equivalencies between subunits. Moreover, the episode points to the familiarity of some students with standard units, which can be important information for planning teaching and learning activities on the topic of length. However, the teacher does not use the student's reference to centimetres to compare the use of standard units and non-standard units. The use of these learning opportunities can be fostered by a deep knowledge of the foundations of length measurement (KoT), such as problematic situations, the comprehension of definitions (standard and non-standard units), certain properties (accumulation and additivity) or procedures (steps in the standard procedure) and some arguments related to the length of measurement. Likewise, KFLM related to the children's use of standard units, the common obstacles in the learning of length measurement and theories of learning length measurement would contribute towards helping the teacher to design her teacher interventions and to use children's references to standard for emphasizing the importance of a consistent unit.

7. Final remarks

The analysis of the episode shows a clear-cut case of a classroom situation in which several intra-mathematical connections sustain a broader extra-mathematical connection. The coordination of the eight connections analysed, and therefore the use of the learning opportunities triggered by them, relies on different kinds of teachers' knowledge being used. This example of an extra-mathematical connection provides insight into the structure of extra-mathematical connections and how they can be established in the classroom.

Understanding extra-mathematical connections as something more than the use of extra-mathematical contexts for the practice of mathematical concepts and skills means considering more specific connections between representations, concepts, properties and procedures that form the complex network of links of the extra-mathematical connection. In this case, the three blocks of connections are related to the foundations of length measurement that are at the core of the activity. Conceptual connections with conversion allow measures made with non-standard units to be translated into numerical values, seeking to show the inverse relationship between the length of the instrument and the numerical outcome. Conceptual connections with treatment are related to the need to move the instrument repeatedly along the object without leaving empty spaces or changing direction, and also to the properties of standard and non-standard units. In addition, the eight connections are related to all the levels in the mathematical complexity of measurement (Table 3.1) as well as to different subdomains of MTSK. As a result, the usefulness of the extra-mathematical connection for the construction of mathematical knowledge depends, at least partially, on the use of intra-mathematical connections for understanding the mathematical foundations of length.

The analysis of these connections shows again how the coordination between the analysis of mathematical complexity and the analysis of the connections in the classroom practice highlights moments in the teaching practice that reveal the need for teachers to be able to identify and interpret the classroom activity so as to make decisions aimed at consolidating key concepts of length measurement at early stages. However, when those intra-mathematical connections are not enhanced, or when they are misused, students' learning opportunities will be underutilized. In three of the intra-mathematical connections presented in the analysis, lost opportunities can be identified. Therefore, the extra-mathematical connection was misused in relation to those learning opportunities, as students' attention was primarily drawn to units and instruments, while some core aspects of measurement, such as equal partition, unit iteration or the relation between non-standard units of measurement and the centimetre, were not addressed.

Although there are several reasons for the teacher to misuse those learning opportunities, e.g. classroom management or her choice of emphasizing the inverse relationship between the length of the instrument and the numerical outcome, knowledge analysis reveals that there are some types of knowledge that would help the teacher to take greater advantage of the learning opportunities arising from the intra-mathematical connections, especially those related to correctness, exactness and reliability.

This knowledge is related to a deep understanding of the mathematical foundations of measurement (KoT). Even if the teacher's main goal was to show the students the need to use standard units, some important issues related to the procedure of length measurement need not have been ignored. Knowledge of the Structure of Mathematics (KSM) is also present when natural and rational numbers are used to represent quantities of length, which is related to the approximate nature of measurements of continuous magnitudes. It is also important for teachers to know the way concepts such as measurement evolve during the school years, as many properties of length measurement are also properties of area or volume measurement (KMLS).

However, a deep understanding of the foundations of length is not sufficient to take advantage of learning opportunities. KFLM, in the context of measurement, is related also to the identification and interpretation of common mistakes such as those related to equal partition and unit iteration. Besides having a deep knowledge of the topics and recognizing some of the potential difficulties students can face when performing measurements of length, it is fundamental to know when is the best moment to use students' ideas and how these ideas can be used and framed to make some important points in the classroom, such as the importance of performing measurement procedures correctly.

It is important to emphasize that the content of all the aforementioned subdomains of knowledge are intertwined, as the MK subdomains (KoT and KSM) are the pillars on which KFLM and KMT are based. Thus, coordination of different kinds of knowledge allows teachers' knowledge to acquire its specialized dimension. The analysis presented in this chapter shows that extra-mathematical connections are based on the coordinated establishment of intra-mathematical connections, triggering learning opportunities that need several types of teacher knowledge to be used.

The analysis leads to a reflection on certain didactic questions related to the use of non-standard units in early years. In the particular case of length measurement, some studies (Clements, 1999) note that the early introduction of non-standard units of measure to show the need to use standard units can be premature. Our results may reinforce this idea, as some students proposed the use of standard units to perform some of the tasks.

Acknowledgement

This research was supported in part by MINECO (Spain) projects EDU2014–54526-R, EDU2015–65378-P, EDU2017–87411-R MINECO/FEDER and SGR-2014–972-GIPEAM. It has also been partially supported by the grant #2016/22557–5, São Paulo Research Foundation (FAPESP), Brazil.

References

Ball, D., Thames, M., & Phelps, G. (2008). Content knowledge for teaching: What makes it special? *Journal of Teacher Education*, *59*(5), 389–407.

Bryman, A. (2004). *Social research methods*. New York: Oxford.

Campbell, N. R. (1928). *An account of the principles of measurement and calculation*. London: Longmans Green.

Carrillo, J., Climent, N., Contreras, L. C., & Muñoz-Catalán, M. C. (2013). Determining specialised knowledge for mathematics teaching. In B. Ubuz, C. Haser, & M. A. Mariotti (Eds.), *Proceedings of the Eighth Congress of the European Society for Research in Mathematics Education* (pp. 2985–2994). Ankara, Turkey: Middle East Technical University and ERME.

Carrillo, J., Climent, N., Montes, M., Contreras, L. C., Flores-Medrano, E., Escudero-Ávila, D. . . . Muñoz-Catalán, M. C. (2018). The Mathematics Teacher's Specialised Knowledge (MTSK) model. *Research in Mathematics Education*. ISSN: 1754–0178 (Online). doi:10.1080/14794802.2018.1479981

Clements, D. H. (1999). Teaching length measurement: Research challenges. *School Science and Mathematics*, *99*(1), 5–11.

Clements, D. H., & Sarama, J. (2007). Early childhood mathematics learning. In F. K. Lester (Ed.), *Second handbook of research on mathematics teaching and learning* (pp. 461–555). Charlotte, NC: Information Age.

De Gamboa, G. (2015). *Aproximación a la relación entre el conocimiento de profesor y el establecimiento de conexiones en el aula* (PhD thesis). Universitat Autònoma de Barcelona, Bellaterra.

De Gamboa, G., & Figueiras, L. (2014). Conexiones en el conocimiento matemático del profesor: propuesta de un modelo de análisis. In M. T. González, M. Codes, D. Arnau, y T. Ortega (Eds.), *Investigación en Educación Matemática XVIII* (pp. 337–344). Salamanca, España: SEIEM.

Duval, R. (2006). A cognitive analysis of problems of comprehension in a learning of mathematics. *Educational Studies in Mathematics*, *61*(1–2), 103–131.

Fonseca, L., & Cunha, E. (2011). Preservice teachers and the learning of geometry. In M. Pytlak, T. Rowland, & E. Swoboda (Eds.), *Proceedings of the Seventh Congress of the European Society for Research in Mathematics Education* (pp. 588–597). Rzeszów, Poland: University of Rzeszów and ERME.

Jakobsen, A., Ribeiro, C. M., & Mellone, M. (2014). Norwegian prospective teachers' MKT when interpreting pupils' productions on a fraction task. *Nordic Studies in Mathematics Education*, 3–4.

Kortenkamp, U., & Rolka, K. (2009). Using technology in the teaching and learning of box plot. In V. Durand-Guerrier, S. Soury-Lavergne, & F. Arzarello (Eds.), *Proceedings of the Sixth Congress of the European Society for Research in Mathematics Education* (pp. 1070–1080). Lyon, France: Institut National de Recherche Pédagogique and ERME.

Lanciano, N. (2004). The processes and difficulties of teachers' trainees in the construction of concepts, and related didactic material, for teaching geometry. In M. A.

Mariotti (Ed.), *Proceedings of the Third Conference of the European Society for Research in Mathematics Education* (pp. 1–10). Bellaria, Italy: University of Pisa and ERME.

Ma, L. (1999). *Knowing and teaching elementary mathematics: Teachers' understanding of fundamental mathematics in China and the United States.* Mahwah, NJ: Lawrence Erlbaum Associates.

Mariotti, M. A. (2011). Proving and proof as an educational task. In M. Pytlak, T. Rowland, & E. Swoboda (Eds.), *Proceedings of the Seventh Congress of the European Society for Research in Mathematics Education* (pp. 61–89). Rzeszów, Poland: University of Rzeszów and ERME.

Meissner, H. (2011). Teaching arithmetic for the needs of the society. In M. Pytlak, T. Rowland, & E. Swoboda (Eds.), *Proceedings of the Seventh Congress of the European Society for Research in Mathematics Education* (pp. 346–356). Rzeszów, Poland: University of Rzeszów and ERME.

Montes, M., Ribeiro, C., Carrillo, C., & Kilpatrick, J. (2016). Understanding mathematics from a higher standpoint as a teacher: An unpacked example. In *Proceedings of the 40th Conference of the International Group for the Psychology of Mathematics Education* (Vol. 3, pp. 315–322). Szedge, Hungrya: PME.

Olivero, F., & Robutti, O. (2001). An exploratory study of students' measurement activity in a dynamic geometry environment. In J. Novotná (Ed.), *Proceedings of the Second Conference of the European Society for Research in Mathematics Education* (pp. 215–226). Mariánské Lázně, Czech Republic: Charles University, Faculty of Education and ERME.

Policastro, M. S., Almeida, A. R., & Ribeiro, M. (2017). Conhecimento especializado revelado por professores da educação infantil e dos anos iniciais no tema de medida de comprimento e sua estimativa. *Revista Espaço Plural, 36*(1), 123–154.

Rafiepour, A., & Karimianzade, A. (2017). Fifth-grade students construct decimal number through measurement activities. In T. Dooley & G. Gueudet (Eds.), *Proceedings of the Tenth Congress of the European Society for Research in Mathematics Education* (pp. 964–971). Dublin, Ireland: DCU Institute of Education & ERME.

Ribeiro, M., Badillo, E., Sánchez-Matamoros, G., & Artès, M. (2016). Discussing a primary prospective teacher practice and analysis on a measurement episode: The role of video analysis. In *40th Conference of the International Group for the Psychology of Mathematics Education-PME.* Szegde, Hungary: PME.

Rondero, C., & Font, V. (2015). Articulación de la complejidad matemática de la media aritmética. *Enseñanza de las Ciencias, 33*(2), 29–49.

Rowland, T., Turner, F., Thwaites, A., & Huckstep, P. (2009). *Developing primary mathematics teaching: Reflecting on practice with the knowledge quartet.* London, UK: Sage Publications.

Sarama, J., Clements, D. H., Barret, J., Van Dine, D. W., & McDonel, J. S. (2011). Evaluation of a learning trajectory for length in the early years. *ZDM-Mathematics Education, 43*, 667–680.

Sensevy, G. (2009). Outline of a joint action theory in didactics. In V. Durand-Guerrier, S. Soury-Lavergne, & F. Arzarello (Eds.), *Proceedings of the Sixth Congress of the European Society for Research in Mathematics Education* (pp. 1643–1653). Lyon, France: Institut National de Recherche Pédagogique and ERME.

Sherin, M. G., Linsenmeier, K., & van Es., E. A. (2009). Selecting video clips to promote mathematics teachers' discussion of student thinking. *Journal of Teacher Education, 60*(3), 213–230.

Shulman, L. S. (1986). Those who understand: Knowledge growth in teaching. *Educational Researcher, 15*(2), 4–14.

Sisman, G. T., & Aksu, M. (2016). A study on sixth grade students' misconceptions and errors in spatial measurement: Length, area, and volume. *International Journal of Science and Mathematics Education*, *14*(7), 1293–1319.

Skemp, R. (1971). *The psychology of learning mathematics*. London, UK: Penguin Books.

Steele, M. D. (2013). Exploring the mathematical knowledge for teaching geometry and measurement through the design and use of rich assessment tasks. *Journal of Mathematics Teacher Education*, *16*(4), 245–268.

Stephan, M., & Clements, D. H. (2003). Linear, area and time measurement in prekindergarten to grade 2. In D. H. Clements & G. Bright (Eds.), *Learning and teaching measurement* (pp. 3–16). Reston, VA, USA: National Council of Teachers of Mathematics.

Subramanian, K. (2014). Prospective secondary mathematics teachers' pedagogical knowledge for teaching the estimation of length measurements. *Journal of Mathematics Teacher Education*, *17*(2), 177–198.

Szilágyi, J., Clements, D. H., & Sarama, J. (2013). Young children's understandings of length measurement: Evaluating a learning trajectory. *Journal for Research in Mathematics Education*, *44*(3), 581–620.

Triantafillou, C., & Potari, D. (2010). Mathematical practices in a technological workplace: The role of tools. *Educational Studies in Mathematics*, *74*(3), 275–294.

Walkerdine, V. (1988). *The mastery of reason: Cognitive developments and the production of rationality*. New York, NY: Routledge.

Zazkis, R., & Mamolo, A. (2011). Reconceptualizing knowledge at the mathematical horizon. *For the Learning of Mathematics*, *31*(2), 8–13.

4

LEARNING TO TEACH TO REASON

Reasoning and proving in mathematics teacher education

*Jeppe Skott, Dorte Moeskær Larsen &
Camilla Hellsten Østergaard*

1. Introduction

Two recurrent recommendations for school mathematics and mathematics teacher education form the background to the pilot study that we present in what follows. First, it is one aspect of current reform efforts in school mathematics that students need to become involved in genuine mathematical activity. Irrespectively of whether this is phrased as process standards (NCTM, 2000), threads of mathematical proficiency (National Research Council, 2001), mathematical practices (Common Core State Standards Initiative, 2010), or competencies (OECD, 2000), it includes an element of mathematical reasoning and proving (R&P). Forms of R&P, then, have been promoted as a significant part of reform efforts for the last couple of decades.

Second, recommendations for mathematics teacher education (MTE) increasingly emphasize issues that are specific to the profession. This is so in suggestions that academic mathematics does not suffice as teachers' content preparation (Ball, Thames & Phelps, 2008; Rowland, Turner, Thwaites & Huckstep, 2009). It is also apparent in the explicit emphasis on the tasks of teaching in the college-based parts of programmes. This involves a shift "from a focus on what teachers know and believe to a greater focus on what teachers do" (Ball & Forzani, 2009, p. 503), and it includes for instance organizing productive classroom communication, selecting and developing challenging tasks and assessing student work for formative purposes. And the professional emphasis in MTE shows in suggestions that teaching-learning processes in MTE should model those envisaged for school mathematics, if reform recommendations are to materialize in school (Lunenberg, Korthagen & Swennen, 2007).

Between them, the two recommendations outlined here indicate not only that R&P should be part of MTE; they also suggest that prospective teachers should work with R&P in ways that resemble the ones recommended for school students.

R&P, however, is notorious for the difficulties it creates for students at all levels, including prospective teachers. In this chapter we present the background and pilot to an intervention study that is to alleviate these problems in the case of Danish teachers for the primary and lower secondary levels. The study, *Reasoning and Proving in Teacher Education* (RaPiTE), is in line with other studies in the field in the sense that we suggest that MTE needs to be close to the challenges of the profession. However, based on previous studies of practising teachers, we suggest not taking this recommendation too far. These studies indicate that there is a need to coordinate the recommendation to professionalize teacher education with a concern for the disciplinary practice of R&P. The rationale of the pilot is to address the questions of whether this dual emphasis on school mathematics and academic mathematics is important also in the education of prospective teachers, and if our interpretation of how such a balance may be conceived is warranted. We address these questions in what follows. The results of the pilot suggest an affirmative answer.

To make our point, we begin by introducing the notion of R&P leading to a presentation of our use of the term. We then discuss recent educational scholarship on R&P in school mathematics and MTE, including the problems that many students face with R&P and suggestions for how to address them. We then present our framework, Patterns of Participation (PoP), and elaborate on the approach we take in RaPiTE, before describing the organization, methods and results of the pilot. We finish with a discussion of our main concern: does it make sense to balance academic and professional issues in mathematics teacher education on R&P, and what may it mean to do so?

2. Reasoning and proving in mathematics education scholarship

The term of mathematical reasoning has been used in different ways, and in spite of the consensus that it is important, there seems to be no common understanding of what it entails, neither in curricular documents nor in the research literature (Jeannotte & Kieran, 2017). One difference is what the concept of mathematical reasoning is meant to encompass. Brousseau and Gibel (2005) define reasoning as a relation between two elements, a condition or observed fact and a consequence. Somewhat similarly, Duval (2007) describes reasoning as "a 'logical linking' of propositions" (p. 140) that may change the epistemic value of a claim from being, for instance, visually obvious to logically necessary or impossible. Duval argues that one of the characteristics of mathematical reasoning is that the truth of a proposition depends on its epistemic necessity. This is unlike other domains in which truth value may be based for instance on perceptual obviousness. From this perspective, mathematical reasoning is a matter of justifying claims by use of deduction.

Others view justifications in the form of proofs as part of a broader concept of reasoning that encompasses also investigating patterns and making conjectures (e.g. A. J. Stylianides & Ball, 2008). This does not do away with the subject-specific, deductive character of mathematical justifications, but it broadens the perspective. NCTM (2008), for instance, locates proof in a reasoning-and-proof cycle of exploration, conjecture and justification. Similarly Gabriel J. Stylianides (2008) focuses on reasoning as a process of making inquiries to formulate a generalization or conjecture and determining its truth value by developing arguments, some of which may qualify as proofs. And in his study of a Grade 5 classroom, Reid (2002) argues that the students are engaged in a process of pattern observation, conjecturing and conjecture testing. In one of the three episodes that he reports on, this leads to generalization and deduction; in another, to refutation of the conjecture; and in the last episode, the students' bar exceptions in order to develop a revised conjecture. While justification of mathematical claims is deductive, these broader R&P processes allow also for inductive and abductive modes of reasoning (Pedemonte & Reid, 2011).

Another contentious issue beyond what the term R&P encompasses is what is required in terms of the style of an argument for it to qualify as proof. In turn this is linked to where the justification is located. Reid & Knipping (2010) argue that to some, the proof is in the proof text, and proving is understood as the writing of a text that is sufficiently similar to what is accepted in academic mathematics. This seems to be what Jeannotte and Kieran (2017) refer to as formal proofs that are "systematized in a mathematical theory" and for which the process of proving "relies on mathematical theory built a priori and on formalized realizations (axioms and theorems)" (p. 13). From this perspective, a generic argument, for instance, does not qualify as (formal) proof. However, as Reid and Vargas (2018) have argued, this is based on an expectation that the proof text explicates the full logic of the argument. However, proof texts are not logically rigorous deductions, but "arguments that such deductions exist" (p. 241). This may also be the case, Reid and Vargas argue, if examples are considered generic and used as the basis for logical deduction, but obviously not if the example is used for an empirical generalization. Also distancing himself from an exclusive focus on formal proving, A. J. Stylianides (2007) suggests that proving in school (and elsewhere) may be understood as making mathematically valid inferences on the basis of what is or may become taken-as-shared in terms of content and modes of argumentation in the community in question.

These latter recommendations seek to engage students at all school levels in a range of R&P processes, including developing deductive mathematical justifications. With regard to justifications, the recommendation is not to treat students as "little mathematicians", but to invite them "to explore or debate the truth of mathematical assertions based on the logical structure of the mathematical system rather than on the authority of the teacher or the textbook" (Gabriel J. Stylianides, A. J. Stylianides & Weber, 2017, p. 237).

Our intention in RaPiTE is similar, as we are interested in developing possibilities for teachers to support such exploration and debate. For our present purposes and in line with the R&P cycle, we use the term mathematical reasoning in educational settings about the dynamic processes of conducting investigations; making, testing and refining conjectures; and developing justifications in the form of deductive arguments based on previously agreed-upon premises or results irrespective of whether the formulation of the argument complies with standard mathematical practice or not. Further, we – in line with others (e.g. A. J. Stylianides, 2007; Gabriel J. Stylianides et al., 2017) – suggest that for an argument to function as proof, it needs not only to be deductive; it is also a requirement that the premises, the modes of reasoning and the language in which it is formulated are accepted by the classroom community.

3. Problems with R&P and suggestions for how to address them

In spite of differences in understanding of what R&P entails, there is some agreement that like other mathematical processes, it is both an educational aim in its own right and a means for the students to develop better understandings of other contents (e.g. Hanna, 2000; Yackel & Hanna, 2003). However, a recurrent theme in the literature is that R&P is treated with less than the care it deserves. Proofs, for instance, are often dealt with in secondary geometry only and exclusively for the purpose of verification of results that are presented ready-made.

As a consequence, a series of interconnected learning problems abound for students at all levels, including teacher education. Students find it difficult to understand why a proof is needed (Zaslavsky & Shir, 2005). Also, mathematical argumentation often degenerates into authoritative or empirical proof schemes (Education Committee of EMS, 2011; Harel, 2007). Further, attempts to engage students in "'reproductive' learning" of proofs are often unsuccessful (Pedemonte, 2007, p. 25), as students may focus on the format rather than on the contents of the argument. This may turn what was to be mathematical reasoning into a ritualistic following of formal procedures. As Brousseau and Gibel say:

> If model proofs are still presented to students, they are meant to serve as "model reasoning" which the students could then use in producing their own original forms of reasoning. But there is always the risk of reducing problem solving to an application of recipes and algorithms, which eliminates the possibility of actual reasoning.
>
> *(Brousseau & Gibel, 2005, p. 14)*

Finally, it is argued that the intention of using R&P to support student learning of other contents (e.g. geometry, algebra, etc.) is often not realized.

The processual approach to R&P in current recommendations for reform (cf. the previous section) may be seen as an attempt to address some of the problems outlined earlier. For instance, it has been argued that there is some continuity between the phases of the R&P cycle. This means that *cognitive unity* may develop between exploring and conjecturing on the one hand and proving on the other, if students are engaged in different R&P processes with suitable support (e.g. Arzarello, Bussi, Leung, Mariotti & Stevenson, 2012). In turn, this may facilitate improved understanding of why a proof is needed and avoid an emphasis on proving as a ritualistic following of procedures.

A related recommendation concerns the intention of using R&P to support students' understanding of the contents in question. Hanna (2000) discusses the functions of proving and says, "the best proof is one that also helps understand the meaning of the theorem being proved: to see not only that it is true, but also why it is true" (p. 8). Hersh (2009) suggests that this is particularly important in educational settings. In research, he says, proofs are to convince, but convincing is all too easily accomplished in the classroom, and the intention is to provide insight as to why a claim is true or false. This means that in school mathematics, the main emphasis should be on *proving why*.

In line with the general trend outlined in the introduction to this chapter, it has been suggested to use approaches to R&P recommended for schools in MTE to address some of the problems mentioned earlier (e.g. Boero, Fenaroli & Guala, 2018). Such suggestions include embedding proving in the broader R&P cycle and focusing more on *proving why*, for instance by using generic arguments (Rowland, 2002).

Our study, RaPiTE, is to some extent in line with this recommendation, but building on studies of practising teachers, we argue that it should not be taken too far. In a sense to be explained later, we suggest that prospective teachers need to become engaged in R&P processes that are "sufficiently close" both to classroom practice and to the discipline of mathematics.

4. Studies of practising teachers

It is, then, one suggestion for how to promote school students' proficiency with mathematical reasoning to engage them in all phases of the R&P cycle and to focus on *proving why*. More specifically, students may build on their own explorations to produce conjectures and develop generic arguments to support them. As an example, consider the case of Larry, a Grade 5 teacher whose class is working on perfect squares and perfect cubes (Skott, 2018a). The students have previously made geometrical representations of square numbers with centicubes (cubes that may be assembled and used e.g. for teaching place value). The class has now made a table of the natural numbers from 1 to 14 and their squares on the board. This leads to the observation that $5^2 - 4^2 = 9 = 5 + 4$. A *single-case key idea inductive argument* (Morris, 2007) to justify the observation might have been built on the geometric representations used before. Placing two squares with side lengths 4 and 5 on top of each other with two pairs of sides aligned provides an explanation of the result and may be used to develop a generic argument that the difference between two consecutive perfect squares is the sum of their bases.

There may, then, be some potential for addressing the problems with R&P in school classrooms in the suggestion to engage students in investigating and conjecturing as an entry to *proving why*. However, this is not easily accomplished. In particular, students may become involved in the first two phases of the reasoning-and-proof cycle but still rely on empirical or other justifications that do not qualify as mathematical. That was often the case in Larry's classroom (cf. the earlier example). For instance, Larry never capitalized on the students' conjecture that $(n + 1)^2 - n^2 = (n + 1) + n$ and sought to develop a mathematically valid justification, generic or otherwise.

Similar situations occur in Bieda's (2010) multiple case study of experienced middle school teachers. The teachers in the study use a textbook that emphasizes R&P, and in class students produce conjectures in response to textbook tasks. However, in only about half the cases do they provide some form of justification, and the students' example-based justifications are accepted as much as their more general ones. The result is that the students have little opportunity to develop understandings of the specifics of mathematical R&P. A possible explanation, Bieda says, is that the teachers become involved in a reform agenda that prioritizes "student-centred teaching", which requires them to play a relatively unobtrusive role in relation to the students' learning.

The studies mentioned here suggest that teachers may find it difficult to capitalize on the R&P potential of situations that "arise naturally from students' work as they explore mathematical phenomena, examine particular cases, discuss alternative hypotheses, and generate conjectures" (A. J. Stylianides & Ball, 2008, p. 312). As a result of the difficulties, mathematical R&P may lose its subject specificity as students and their teachers continue to rely on empirical or other justifications that do not qualify as mathematical. We suggest that the PoP framework may be able to explain why, and we let these explanations inform our development initiatives on R&P in MTE.

5. Framework and rationale of RaPiTE

It is a recurrent explanation of school students' lack of proficiency with R&P that their teachers have limited knowledge of these processes and that their beliefs run counter to current reform efforts. This somewhat acquisitionist perspective primarily locates the educational problems of R&P within the individual teacher. This is so even when it is acknowledged that these individual constructs may be challenged by dominant norms or practices at the schools where prospective teachers do their practicum or where novices take up teaching (e.g. Gabriel J. Stylianides, Stylianides & Shilling-Traina, 2013).

In RaPiTE we adopt a somewhat more participatory stance to human functioning. The framework that we use, PoP, draws on social practice theory (e.g. Holland, Skinner, Lachicotte & Cain, 1998; Lave, 1997; Wenger, 1998), symbolic interactionism (Blumer, 1969; Mead, 1934) and Sfard's theory of commognition (Sfard, 2008). The first and the last of these approaches focus, respectively, on emerging social processes (e.g. romance at a US university campus, cf. Holland & Eisenhart, 1990) and on well-structured cultural practices (e.g. mathematics, cf.

Sfard, 2008). PoP, however, does not focus on any one such practice per se and on how an individual moves towards more comprehensive participation in it. Rather, PoP re-centres the individual and asks how a teacher's involvement in unfolding school and classroom events relates to and is transformed by her or his re-engagement in other past and present practices and discourses (Skott, 2018b). We have found the I-me distinction in the symbolic interactionist notion of self helpful for this purpose. The distinction points to how a person in an interaction acts (does, says, thinks, etc.); that is, the person performs as an I but instantaneously adjusts the act as (s)he takes the attitude to her- or himself of others and becomes a me. The attitude (s)he takes to her- or himself may be that of immediate interlocutors or of significant individual or generalized others that come to mind at the instant. The distinction between the I and the me allows us to focus on how the teacher may see her- or himself from the perspective of the teacher's students or other individual and generalized others as classroom processes unfold.

If, for instance, a teacher seeks to develop a good mathematical argument with a group of students, who appear to be weak and vulnerable in the situation, (s)he may simultaneously take the attitude to her- or himself of the students in question; of colleagues, who focus on creating trusting relationships with the students; of the school leadership or of parents, who emphasize students' performance on standardized tests; or of her or his teacher education programme that focuses on the use of manipulatives to facilitate student learning with understanding (Skott, 2013, 2015; Skott, Larsen & Østergaard, 2011). The teacher's engagement with each of these social constellations – or others – may transform or subsume her or his involvement in the practice of mathematical R&P and, for instance, have the teacher accept justifications that do not qualify as mathematical. PoP provides a perspective on if and how this is the case.

PoP has so far framed studies conducted "in the perspective of teacher education" (Krainer & Goffree, 1998). These studies are not on MTE, but they develop understandings of teaching-learning practices in schools and may raise questions about MTE and inform decisions on how to address them. As indicated in the previous section, the results suggest that even when teachers engage students in elements of the R&P cycle, most notably what NCTM (2000) calls "examining patterns and noting regularities" (p. 262), modes of justification may lose their subject specificity. To avoid this, it seems that MTE needs to fulfil two requirements. First, it must be close to teaching-learning processes in schools, as the community in which R&P practices develop are otherwise too distant from classroom interaction for it to function as a generalized other in instruction. This is in line with the suggestion to emphasize *proving why* using generic arguments (Rowland, 2002). Second, and in spite of that, MTE must be close to the disciplinary practice of R&P and include significant elements of *proving that* so as to limit the risk of classroom processes losing their subject specificity. This is important also because it is often more complicated to *prove why* than to *prove that*, and if the teacher is unable to produce a valid mathematical justification in response to an unforeseen student conjecture, (s)he has no alternative but to accept the students' empirical arguments. The assumption of RaPiTE, then, is that MTE needs to avoid the two extremes of focusing either on academic mathematics or school mathematics, not by reducing the emphasis on either but by transforming both (Skott, 2018a).

To be "sufficiently close" to both school mathematics and academic mathematics, we use tasks and conjectures that may be used in or developed from tasks used in school and take them beyond the level of the school students in question. Examples include:

1   Does 8 always divide $n^2 - 1$, if n is odd? (From an interaction in Larry's Grade 5; cf. the previous example on perfect squares, Skott, 2018a)
2    $30 = 6 + 7 + 8 + 9$; $31 = 15 + 16$; $32 = ...$; $35 = 17 + 18 = 5 + 6 + 7 + 8 + 9 = ...$; What positive integers are the sum of other consecutive, positive integers? (Arose from a reversal of the question of how to find the sum of the first n positive integers)
3   Assume that you have a set of rods similar to Cuisenaire rods representing the positive integers from 1 to n. For what values of n can you make two "trains" of rods of equal length? Three trains? m trains?

6. The pilot study

The pilot study takes place at a prestigious college in Denmark. The prospective teachers have all performed fairly well in secondary school, and according to curricular documents they have worked with mathematical reasoning both in primary and secondary school. At the college, they need to specialize in teaching either Grades 1–6 or 4–9, and in teaching either Danish or mathematics. All research participants in RaPiTE are to teach Grades 4–9, and they are all among the 35% of the prospective teachers (from now on, teachers), who specialize in mathematics.

There are two phases in the pilot study. The first phase consists of a questionnaire that the prospective teachers received on the day of their first mathematics class at the college (for more detail, see the next section). The second phase, which is the focus of the present chapter, consists of three parts. First, the participants attended a short teaching-learning sequence on R&P at the college; second they were expected to work with R&P with their students in their first practicum; and third, they were to present and analyse video clips on R&P from their practicum when back at the college. Fifty-seven prospective teachers participated in the first phase, 31 in the second.

Phase 2 of RaPiTE was included as part of the first of four compulsory courses on mathematics and mathematics education for teachers specializing in the subject. The bulk of the course is on numbers and algebra and on students' learning of these topics. The teachers are expected to spend 275 working hours on the course, including 70 lessons in classrooms with a teacher educator. The practicum is not normally linked to the course in mathematics, and the learning goals for the practicum are phrased in general terms concerning, for instance, class-room organization and building and maintaining productive relationships with students. In RaPiTE we sought to bridge the usual divide between the math-ematics course and the practicum.

6.1 Organization and methods

The two phases of the pilot study provide very different settings for the par-ticipants, and we do not expect the questionnaire from the first phase and the observations of the teaching-learning sequence from the second to shed light on relatively stable and context-independent mental constructs. Also, we do not assume any causal relation between responses to the questionnaire and teachers' contributions to classroom practice. At best, the questionnaire allows us to understand how the teachers react discursively to R&P in a set-ting in which they are not challenged by other concerns that may emerge in classroom interaction. From a PoP perspective, it is an empirical question whether teachers orient themselves towards such a discourse as they engage with their students in the classroom. However, if teachers face significant problems with R&P in the questionnaire, we consider it unlikely that they engage proficiently with these processes when teaching. In other terms, it seems a necessary but by no means a sufficient condition for their proficient engagement with R&P in the classroom that they are able to deal with these processes in the questionnaire.

As mentioned earlier, the research participants filled out a questionnaire at the very beginning of the academic year. The questionnaire consists of open items on why they decided to go into teaching, why they chose to specialize in math-ematics and what their general experiences are with school mathematics. They are also asked about specific experiences with R&P (e.g. "Describe how you felt about reasoning and proofs in mathematics") and to comment on situations from school mathematics with an element of mathematical reasoning, including situations with students working with R&P (Larsen, Østergaard & Skott, 2018).

As mentioned earlier, the present chapter focuses on the second phase of the pilot. The first part of this phase is a 12-lesson teaching-learning sequence on R&P organized as two sessions of six 45-minute lessons. This sequence was not taught by the authors of this chapter, but the second and third authors planned it and developed the teaching-learning materials. The intentions and the contents were discussed in detail with the colleague, who taught the sequence.

In the sequence the teachers are introduced to R&P and to different argu-ments for why to engage with R&P in school mathematics. More specifically, it was discussed why R&P is important in its own right and what one may expect in terms of student learning; how R&P may support the students' understanding of other contents; what the character, advantages and disadvantages are of dif-ferent types of argumentation, including possible transitions from empirical to deductive reasoning; and what the differences are between definitions, axioms and theorems in mathematics. Also, it was discussed how one may seek to create learning environments that support students' learning of R&P. This discussion was based, for instance, on a video recording of school students making and justifying conjectures about a number pattern in a sequence of geometric figures. This leads to discussions about the quality of the students' arguments and how students may be supported in developing them further. Subsequently, the teachers become involved in all three parts of the R&P cycle, for instance as they work on a version of the third task mentioned previously on making "trains" of equal length out of Cuisenaire rods (cf. section 5). As part of this, they are to make geometrical or number theoretical justifications for their claims. Finally the teachers discuss comments from school students, who have previously worked on the same task. One of these reads:

> If I am to make two trains of equal length the sum must be even. If the sum is odd, I would have one left over. If the sum is even there could be other problems. . . . We do not know if it is sufficient that the sum is even.

After these sessions on mathematical R&P, the teachers form eight groups of three or four, each group going on a two-week practicum in a middle or lower secondary school (Grades 4–6 and 7–9, respectively). In the practicum each group normally teaches 8–10 lessons of mathematics a week as well as other subjects for approximately 4–6 lessons. As part of the pilot study, the students are before and during their practicum to (1) plan for their students' involvement in R&P; (2) video record each other's teaching and (3) select one video clip from the practicum in which the students are particularly involved in R&P. After the practicum, and as the last part of the second phase of the pilot study, the teachers discuss the video clips and the inherent potentials for and problems with R&P in a whole-day session. In what follows, we focus on the teachers' response to the last requirement and on the subsequent discussion.

The sequences on R&P at the college before and after the practicum (parts 1 and 3 of the second phase) were video-recorded and transcribed. Like the responses to the questionnaire, the transcripts were analysed with no pre-developed set of codes, using coding procedures inspired by grounded theory (Charmaz, 2006). The initial coding and categorization of the data was first done independently by the second and third authors. The coding included word-by-word, line-by-line, incident-to-incident and in Vivo coding, and memo-writing was used in all phases. Subsequently, codes and categories were compared and discussed among all authors, and inconsistencies were resolved.

The analysis resulted in categories on (1) teachers' reasons for selecting the specific video clips; (2) the character of R&P in the teachers' discussions of these processes; (3) school students' learning of R&P and their learning of other contents from R&P; (4) the general significance of R&P in school mathematics and (5) "blackboard-talk", that is, whole-class teaching as it relates to student learning in general and to R&P in particular.

6.2 Results

The results from the first phase of the study, the questionnaire, support previous findings that many teachers have difficulties with deciding what a valid mathematical argument is. When asked to comment on a student's false conjecture about a connection between the perimeter and area of a rectangle, 26 of the participants provided an acceptable answer and one did not respond to the item. 30 participants accepted the student's claim without questioning its empirical base.

Fifty-four participants responded to another questionnaire task in which they were to explain why the sum of two odd numbers is even. Twelve provided an acceptable or almost acceptable answer, some of them phrasing it with little use of mathematical symbolism (e.g. "an odd number has a remainder of 1 if divided by 2. Two odd numbers with remainder 1 that are added have the remainder 2 → no remainder"). Among the unacceptable answers, there are 11 empirical arguments (e.g. "One can see this, if you just try a sufficient number of times"; "1 + 1 = 2, 3 + 1 = 4, 1 + 5 = 6, 1 + 7 = 8, . . .").

More surprisingly, there is little connection between the teachers' affective relation with mathematics and their assessment of their own mathematical qualifications on the one hand, and their proficiency with R&P on the other (Larsen et al., 2018). There are 46 research participants, who consider themselves highly qualified in mathematics and/or who are fond of the subject. Only 18 of these participants provide an (almost) acceptable response to the perimeter-and-area item, and 8 do so to the item on the sum of two odd numbers.

In the observations from the college classroom in the last part of the second phase of the study, the teachers face considerable problems arguing how or why the video clips they selected from their practicum are related to R&P. Four of the groups do not provide a coherent explanation for why they selected the episode, and three of the other groups have selected their clips for reasons that are unrelated to mathematical reasoning (e.g. the teachers' supervisor chose the video clip or the technical quality of the recording was good). The last group claims that their clip is on reasoning, but it shows students making number stories for tasks on fractions.

There may, of course, be many reasons why the prospective teachers brought video clips back from their practicum with little or no connection to R&P. For instance, they may have focused on more general educational problems, or their students may have found R&P difficult and for that reason refrained from engaging in these processes. However, there is nothing in what the prospective teachers said or did when back at the college that indicates that any of this was the case, and they in no way suggested that they were not satisfied with their selection of the clips as a reasonable response to the task that had been set before their practicum. We therefore suggest that their selection is based on insufficient prior experiences with R&P.

Looking at the clips themselves, rather than at the teachers' reasons for selecting them, three have no connection to mathematical reasoning (e.g. the teacher presents the solution to a procedural task on the board). Other episodes have some potential for student involvement in R&P, but the teachers do not emphasize aspects of R&P in the discussion with their students in the classroom.

In one episode with some potential for student involvement in R&P, middle school students are to find the point equidistant from the vertices of a triangle. In the video clip, a school mentor, the teacher normally teaching the class, unintentionally shows the students an incorrect procedure for constructing perpendicular bisectors. The students use the incorrect procedure, but having measured the distances on their drawings, they realize that something is wrong. They then shift their attention to the question of how to draw a perpendicular bisector, but they pay no attention to how and why it may help them solve the initial problem.

In their discussion of the video clip at the college, the teachers focus on what they describe as lack of conceptual understanding on the part of the students and on what they appear to consider general drawbacks of whole-class instruction. There is no discussion of if and how the task may become a point of departure for the students' exploration of the problem or for formulating and justifying a conjecture about the properties of perpendicular bisectors.

Another episode with some potential for involving the students in the phases of the R&P cycle concerns the pattern in the number of squares in a sequence of figures. The teachers introduce the task and show the two first figures in the sequence on the board (Figure 4.1). The students, who are in Grade 6, use small cookies to represent the squares. The video clip shows two students, who have 33 cookies. They have written "7" and "y = 4x + 5" and made the drawing in Figure 4.2.

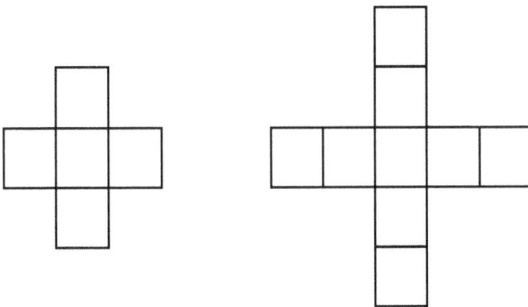

FIGURE 4.1 The two first figures in the sequence as shown on the board.

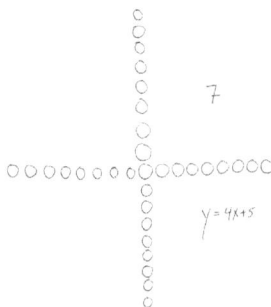

FIGURE 4.2 The group's drawing of their cookies.

The communication between the teacher and the students in the video clip is as follows:

TEACHER: The equation, you are working with, what is x in that equation? I can see you have written 7, but what is the x in your figure? If you look at the cookies, what is the x? . . .

ANNA: If we count. . . . Those five cookies, you don't count them [points at the five cookies in the middle]. You only count these [points at the "arms"], don't you?

TEACHER: The figure that was on the board, how about it?

ANNA: Do you count these [the cookies in the middle] and add them to the "arms"?

TEACHER: In the beginning there were some cookies, right? What do you think are going to be the "arms"?

ANNA: How do you make the "arms"? Do you take that cookie [points at one of the cookies next to the one in the middle]? I think you count how long your "arms" are.

TEACHER: Is this cookie a part of the "arm"? [Points at one of the cookies next to the one in the middle.]

ANNA: Yes, I think so, right?

TEACHER: How can we characterize the "arms", if we want this cookie to be included [points at the cookie next to the one in the middle]?

ANNA: Then these are the "arms" [points at a row of cookies, leaving out only the one in the middle], the rest of the cookies belong to the "arms" [moves the cookies around]. But this is not right!

BENNY: I think there are too many cookies, right? [Counts the cookies in the "arms".]

TEACHER: It looks as if you have counted them perfectly . . .

The students have trouble linking the equation with the geometric representation they made themselves, and as the teacher joins them, the emphasis of the discussion is on the meaning of x and the length of the arms of the cross. These difficulties appear also in other groups, who have written the same equation.

Back at the college, the teachers' discussion of the episode tends to revolve around general pedagogical issues. Major concerns are student motivation; how it may become a successful experience for the students, "success" understood in terms of the students being comfortable and having a good time; and how much time students should spend on working on their own before receiving help from the teacher. The discussion does not focus on the potential of the episode as an opportunity for the students' learning of R&P.

Issues related to the contents of the episode do come up in the discussion. The teachers focus primarily on the students' problems with linking the sequence of figures to the symbolic representation in the equation, and on their difficulties deciding on the value of the constant, b, in the equation $y = mx + b$:

TEACHER (from the group that selected the episode): They [the students who responded to the task] begin by showing the 5 cookies in the middle. And that's what confuses the students. After that they write $y = 4x + 5$ instead of $4x + 1$, and where the first cookie in the arm is included. And that's their problem. . . .

TEACHER (from the group that selected the episode): [T]hey have no problem making the additions . . . and finding rules or patterns. The problem appears when they are to make the equation. And that's what we want them to make – an equation. And here they also have some problems with symbolic understanding.

Another teacher suggests that there may be some potential in discussing how the number of squares in each figure may be calculated from the number of squares in the previous one:

TEACHER (not from the group that selected the episode): It would be nice with a table and count down to one [that is, to the first figure in the sequence]. And then they have a function that says that four will be added every time.

The teachers do not discuss if the equation could be considered a conjecture that could be tested and revised or verified. In this sense, mathematical justifications do not become an issue.

7. Discussion and conclusions

Much of the reform literature seeks a balance in school mathematics as well as in MTE that does justice to the disciplinary practice of R&P, while also paying significant attention to the students as learners of the subject (e.g. A. J. Stylianides & Ball, 2008). It is certainly part of the background to RaPiTE to promote such a balance in relation to mathematical reasoning. The question is what it entails in MTE.

The pilot confirms that teachers often face problems with R&P. The questionnaire establishes a setting remote from the classroom, and the results do not in and by themselves indicate how the teachers react to similar questions when teaching. However, a PoP perspective suggests that the risk of not engaging sufficiently with R&P is greater in classrooms with many other pressing concerns beyond the quality of a mathematical argument.

The second phase of the pilot is more closely related to classroom teaching. The observations suggest that the teachers are concerned with general pedagogy to the extent that they lose sight of content matter issues, including R&P. This is so even when they have been asked to focus specifically on mathematical reasoning. Also, the teachers face problems identifying classroom situations with a potential for student involvement in R&P, and in episodes with some such potential the modes of reasoning tend to lose their subject specificity.

In the episode with students seeking to find a point that is equidistant from the vertices of a triangle, it does not become a topic of discussion in the classroom whether there is such a point and why perpendicular bisectors may solve the problem. Also, when back at the college, the teachers do not focus on how students may become involved in the types of reasoning required to justify that the perpendicular bisectors do indeed provide a solution.

In the cookie episode, the students engage in the important task of finding and generalizing a pattern. The students have written an equation that does generalize it, but it is not clear from the video whether their argument is empirical, in which case they have just found a pattern in the numbers, a pattern that still needs to be confirmed. The alternative is that they have actually produced a generic argument that shows why the equation is right for all values of x. The generic argument is certainly within the reach of the students in Grade 6: *there are 7 cookies in each arm and 4 arms; 7 × 4 and adding the 5 cookies in the middle gives you 33. That argument would be the same for 6, 8, or any other number of cookies in each arm.*

It is hardly surprising that the students do not by themselves explicate the generic argument. It may be more surprising that the teachers do not engage with the question of the character of the students' reasoning, either when with the students in classroom or in discussions of the episode at the college.

In this case a generic argument explains *that* and *why* the equation y = 4x + 5 (or y = 4x + 1) generalizes the pattern. As with many other patterns, however, the task also invites an argument based on mathematical induction. We consider it worth noting that none of the teachers considered informal mathematical induction an option in the discussion at the college, for instance phrased like this: *we have seen that y = 4x + 5 for at least some x (e.g. x = 7). In every new figure the number of cookies increases by 4, so that for the next figure we get y = 4x + 5 + 4 = 4(x + 1) + 5.*

In other situations, students' suggestions do not lend themselves as easily to generic arguments that *prove why*. This is the case for instance with the conjecture from Larry's classroom (cf. section 5) that if n is odd, 8 divides $n^2 - 1$. Straightforward algebraic arguments and proof by induction may be used to prove the conjecture. However, if teachers have not engaged sufficiently with such arguments to have a sense of why they are needed, and if they are not sufficiently proficient with using them, they have no alternative but to rely on the empirical observations of the students to confirm the conjecture.

From a PoP perspective, these examples indicate that teachers need significant experiences with both *proving that* and *proving why*, if they are to support R&P activities in the classroom. The emphasis on *proving that* does not advocate a return in MTE to standard university courses that have no relation to classroom practice. The mathematical practices involved are in that case too remote from school mathematics for teachers to draw on them in classroom interaction. However, using examples from school mathematics to develop means of *proving that*, including the much-criticized proof by induction (Rowland, 2002), is necessary if teachers are to develop sufficient proficiency with dealing with all aspects of the R&P cycle in the classroom and capitalize on the potentials of the students' conjectures.

The pilot study examined the feasibility of proposals for MTE that are based on research with practising teachers. These studies suggest that MTE needs to be close to both school and academic mathematics for teachers to link their mathematical proficiency to instruction. In the case of R&P, this means developing genuinely mathematical modes of justification in relation to tasks and conjectures that may be used in or developed from classroom interaction. The pilot confirms that this may be one way of providing teachers with some of the experiences with R&P they need to support their students' learning of these processes.

References

Arzarello, F., Bussi, M. G. B., Leung, A. Y., Mariotti, M. A., & Stevenson, I. (2012). Experimental approaches to theoretical thinking: Artefacts and proofs. In G. Hanna & M. d. Villiers (Eds.), *Proof and proving in mathematics education* (pp. 97–137). Dordrecht: Springer.

Ball, D. L., & Forzani, F. M. (2009). The work of teaching and the challenge for teacher education. *Journal of Teacher Education*, *60*(5), 497–511. doi:10.1177/0022487109348479

Ball, D. L., Thames, M. H., & Phelps, G. (2008). Content knowledge for teaching: What makes it special? *Journal of Teacher Education*, *59*(5), 389–407. doi:10.1177/0022487108324554

Bieda, K. N. (2010). Enacting proof-related tasks in middle school mathematics: Challenges and opportunities. *Journal for Research in Mathematics Education*, *41*(4), 351–382.

Blumer, H. (1969). *Symbolic interactionism: Perspective and method*. Berkeley: University of Los Angeles Press.

Boero, P., Fenaroli, G., & Guala, E. (2018). Mathematical argumentation in elementary teacher education: The key role of the cultural analysis of the content. In A. J. Stylianides & G. Harel (Eds.), *Advances in mathematics education research on proof and proving: An international perspective* (pp. 49–67). Cham: Springer.

Brousseau, G., & Gibel, P. (2005). Didactical handling of students' reasoning processes in problem solving situations. *Educational Studies in Mathematics*, 13–58. doi:10.1007/0-387-30451-7_2

Charmaz, K. (2006). *Constructing grounded theory: A practical guide through qualitative analysis*. London, UK: Sage Publications.

Common Core State Standards Initiative. (2010). *Common core state standards for mathematics*. Retrieved from www.corestandards.org/wp-content/uploads/Math_Standards.pdf

Duval, R. (2007). Cognitive functioning and the understanding of mathematical processes of proof. In P. Boero (Ed.), *Theorems in schools: From history, epistemology and cognition to classroom practice* (pp. 137–161). Rotterdam, The Netherlands & Taipei: Sense Publishers.

Education Committee of EMS. (2011). *Do theorems admit exceptions? Solid findings in mathematics education on empirical proof schemes*. Retrieved from www.euro-math-soc.eu/ems_education/education_homepage.html#reports

Hanna, G. (2000). Proof, explanation and exploration: An overview. *Educational Studies in Mathematics*, *44*(1), 5–23.

Harel, G. (2007). Students' proof schemes revisited. In P. Boero (Ed.), *Theorems in school: From history, epistemology and cognition to classroom practice* (pp. 65–78). Rotterdam, The Netherlands: Sense Publishers.

Hersh, R. (2009). What I would like my students to already know about proof. In D. A. Stylianou, M. Blanton, & E. J. Knuth (Eds.), *Teaching and learning proof across the grades: A K-16 perspective* (pp. 17–20). New York & London: Routledge & NCTM.

Holland, D., Skinner, D., Lachicotte, W., Jr, & Cain, C. (1998). *Identity and agency in cultural worlds.* Cambridge, MA: Harvard University Press.

Holland, D. C., & Eisenhart, M. A. (1990). *Educated in romance. Woman, achievement and college culture.* Chicago, London: University of Chicago Press.

Jeannotte, D., & Kieran, C. (2017). A conceptual model of mathematical reasoning for school mathematics. *Educational Studies in Mathematics, 96*(1), 1–16. doi:10.1007/s10649-017-9761-8

Krainer, K., & Goffree, F. (1998). Investigations into teacher education: Trends, future research, and collaboration. Paper presented at *the CERME1*, Osnabrück.

Larsen, D. M., Østergaard, C. H., & Skott, J. (2018). Prospective teachers' approach to reasoning and proof: Affective and cognitive issues. In H. Palmér & J. Skott (Eds.), *Students' and teachers' values, attitudes, feelings and beliefs in mathematics classrooms: Selected papers from the 22nd MAVI conference* (pp. 53–63). Cham, Switzerland: Springer.

Lave, J. (1997). The culture of acquisition and the practice of learning. In D. Kirshner & J. A. Whitson (Eds.), *Situated cognition: Social, semiotic, and psychological perspectives* (pp. 17–35). Mahwah, NJ: Lawrence Erlbaum Associates.

Lunenberg, M., Korthagen, F., & Swennen, A. (2007). The teacher educator as a role model. *Teaching and Teacher Education, 23*(5), 586–601. doi:10.1016/j.tate.2006.11.001

Mead, G. H. (1934). *Mind, self, and society from the standpoint of a social behaviorist* (C. W. Morris, Ed.). Chicago: University of Chicago Press.

Morris, A. K. (2007). Factors affecting pre-service teachers' evaluations of the validity of students' mathematical arguments in classroom contexts. *Cognition and Instruction, 25*(4), 479–522.

National Research Council. (2001). *Adding it up: Helping children learn mathematics.* Washington, DC: National Academy Press.

NCTM. (2000). *The principles and standards for school mathematics.* Reston, VA, USA: National Council of Teachers of Mathematics.

NCTM. (2008). *Navigating through reasoning and proof in Grades 9–12.* Reston, VA, USA: National Council of Teachers of Mathematics.

OECD. (2000). *Measuring student knowledge and skills: The PISA 2000 assessment of reading, mathematical and scientific literacy.* Paris: OECD Publishing.

Pedemonte, B. (2007). How can the relationship between argumentation and proof be analysed? *Educational Studies in Mathematics, 66*(1), 23–41. doi:10.1007/s10649-006-9057-x

Pedemonte, B., & Reid, D. (2011). The role of abduction in proving processes. *Educational Studies in Mathematics, 76*(3), 281–303. doi:10.1007/s10649-010-9275-0

Reid, D. A. (2002). Conjectures and refutations in grade 5 mathematics. *Journal for Research in Mathematics Education, 33*(1), 5–29. doi:10.2307/749867

Reid, D. A., & Knipping, C. (2010). *Proof in mathematics education: Research, learning and teaching.* Rotterdam, The Netherlands: Sense Publishers.

Reid, D. A., & Vargas, E. V. (2018). When is a generic argument a proof? In A. J. Stylianides & G. Harel (Eds.), *Advances in mathermatics education research on proof and proving: An international perspective* (pp. 239–252). Cham, Switzerland: Springer.

Rowland, T. (2002). Generic proofs in number theory. In S. R. Campbell & R. Zazkis (Eds.), *Teaching and learning number theory* (pp. 157–184). Westport: Ablex.

Rowland, T., Turner, F., Thwaites, A., & Huckstep, P. (2009). *Developing primary mathematics teaching: Reflecting on practice with the knowledge quartet.* Los Angeles: Sage Publications.

Sfard, A. (2008). *Thinking as communicating: Human development, the growth of discourses, and mathematizing.* Cambridge, UK: Cambridge University Press.

Skott, J. (2013). Understanding the role of the teacher in emerging classroom practices: Searching for patterns of participation. *ZDM: The International Journal on Mathematics Education, 45*(4), 547–559. doi:10.1007/s11858-013-0500-z

Skott, J. (2015). Towards a participatory approach to "beliefs" in mathematics education. In B. Pepin & B. Rösken (Eds.), *From beliefs to dynamic affect systems in mathematics education: Exploring a mosaic of relationships and interactions* (pp. 3–23). Cham, Switzerland: Springer.

Skott, J. (2018a). Goldilocks principle revisited: Understanding and supporting teachers' proficiency with reasoning and proof. In P. Błaszczyk & B. Pieronkiewicz (Eds.), *Mathematical transgressions* (pp. 151–166). Kraków, Poland: Universitas.

Skott, J. (2018b). Re-centring the individual in participatory accounts of professional identity. In G. Kaiser, H. Forgasz, M. Graven, A. Kuzniak, E. Simmt, & B. Xu (Eds.), *Invited lectures from the 13th International Congress on Mathematical Education* (pp. 601–618). Cham, Switzerland: Springer.

Skott, J., Larsen, D. M., & Østergaard, C. H. (2011). From beliefs to patterns of participation: Shifting the research perspective on teachers. *Nordic Studies in Mathematics Education, 16*(1–2), 29–55.

Stylianides, A. J. (2007). Proof and proving in school mathematics. *Journal for Research in Mathematics Education, 38*(3), 289–321.

Stylianides, A. J., & Ball, D. L. (2008). Understanding and describing mathematical knowledge for teaching: Knowledge about proof for engaging students in the activity of proving. *Journal of Mathematics Teacher Education, 11*(4), 307–332. doi:10.1007/s10857-008-9077-9

Stylianides, G. J. (2008). An analytic framework of reasoning-and-proving. *For the Learning of Mathematics, 28*(1), 9–16.

Stylianides, G. J., Stylianides, A. J., & Shilling-Traina, L. (2013). Prospective teaches' challenges in teaching reasoning-and-proving. *International Journal of Science and Mathematics Education, 11*(6), 1463–1490.

Stylianides, G. J., Stylianides, A. J., & Weber, K. (2017). Research on the teaching and learning of proof: Taking stock and moving forward. In J. Cai (Ed.), *Compendium for research on mathematics* (pp. 237–266). Reston, VA, USA: National Council of Teachers of Mathematics.

Wenger, E. (1998). *Communities of practice: Learning, meaning, and identity.* Cambridge, UK: Cambridge University Press.

Yackel, E., & Hanna, G. (2003). Reasoning and proof. In J. Kilpatrick, W. G. Martin, & D. Schifter (Eds.), *A research companion to "principles and standards for school mathematics"* (pp. 227–236). Reston, VA, USA: National Council of Teachers of Mathematics.

Zaslavsky, O., & Shir, K. (2005). Students' conceptions of a mathematical definition. *Journal for Research in Mathematics Education,* 317–346.

5

THE ROLE OF FRAMEWORKS IN RESEARCHING KNOWLEDGE AND PRACTICES OF MATHEMATICS TEACHERS AND TEACHER EDUCATORS

Ronnie Karsenty

1. Introduction

1.1. Rationale and scope of this chapter

As with any other research domain, research in mathematics education can be viewed as accumulating through trends, resulting from the recognition of important issues for which the community has yet to develop deep understanding. Often in the midst of such trends, towards what will later appear as their apogee, or somewhat afterwards, scholars are able to identify the trend and the avenues it took, review the progress made and suggest future directions. In a paper published over a decade ago, da Ponte and Chapman (2006) reviewed a research trend that may be named "mathematics teachers' knowledge and practices", as the title of the paper suggest. They analysed this trend by considering contributions made to PME conferences throughout three decades, and they concluded that

> research of teacher knowledge and practices has made extraordinary progress in these 29 years of PME conferences. We expect this progress to continue, but with some shifts in focus that could take us to new ways or levels of understanding the mathematics teacher and the teaching of mathematics.
>
> *(da Ponte & Chapman, 2006, p. 488)*

Specifically, da Ponte and Chapman point to "the need for reconsidering theoretical and methodological orientations . . . the development or adaptation of innovative research designs to deal with the complex relationships among various variables, situations or circumstances that define teachers' activities" (ibid).

Looking back, it can be said that the research domain of mathematics teachers' knowledge and practices has further developed since the publication of da Ponte and Chapman's review, and indeed much progress has been made in terms of theoretical perspectives. From frameworks such as MKT (Mathematical Knowledge for Teaching; Ball, Thames & Phelps, 2008), and the more recent model of MTSK (Mathematics Teacher's Specialized Knowledge; Carrillo et al., 2018),[1] created for unpacking mathematics teachers' knowledge, to frameworks such as TRU (Teaching for Robust Understanding; Schoenfeld, 2018), created for unpacking mathematics teachers' practices and characterizing the quality of mathematics instruction, the field has known a considerable growth which broadened the community's understanding of what it means to be a mathematics teacher.

What has also happened in the past decade is the rise of another research trend: knowledge and practices of mathematics teacher educators (e.g. Even, 2008; Even & Krainer, 2014; Goos, 2009; Goos & Beswick, in press; Jaworski & Wood, 2008; Prediger, Roesken-Winter & Leuders, 2019). Judging from the recent opening of conference groups dedicated to this topic, in addition to the specialized ETE (Educating the Educators) conference that began in 2014, and several Special Issues in leading mathematics education journals (e.g. Jaworski & Huang, 2014), it seems that this trend's apogee is still ahead of us. Within this relatively new domain, the accumulating research is also expected to develop some theoretical perspectives.

As a step forward towards this aim, and in a similar way to the development at the teacher level, advancement often involves the search for appropriate frameworks to work with. In this chapter, I explore the notion of *framework* and its utilization for conceptualizing our understanding of knowledge and practices of mathematics teachers and mathematics teacher educators. I begin with defining the term *framework*, followed by a brief account on different ways by which frameworks may be created. Then, I focus on what I refer to as a *double-level use of frameworks*, i.e. frameworks that can serve for both researching teachers and researching teacher educators. Special attention is given within this chapter to adapting frameworks from the classroom level to the professional development (PD) level. Finally, I describe in some detail a specific case of such a process of framework adaptation.

1.2. What is a framework?

Although the term *framework* appears extensively in research papers, it is often used in idiosyncratic ways; different researchers mean very different things when referring to frameworks, and specifically so when it comes to *conceptual frameworks* (Ravitch & Riggan, 2016). Thus, I explain herein what I mean by *framework* in the context of this chapter, on the basis of several existing definitions.

Dictionary definitions of the word framework (e.g. Cambridge, Webster) generally refer to *a system or a structure that connects ideas, concepts, beliefs, etc., in order to present a plan or reach a decision*. Such depiction does not seem to align with the context of research, which attempts to understand and explain phenomena rather than present a plan or arrive at a decision; however, what can be borrowed from these definitions is the basic idea that a framework is essentially a multi-component structure, where connections among the different elements shed light on the "big picture" as a whole. This observation is in agreement with several definitions of framework within educational and social science research. For example, Maxwell (2005) defines a conceptual framework as the system of concepts, assumptions, expectations, beliefs and theories that supports and informs research. Imenda (2014) defines a conceptual framework as "an end result of bringing together a number of related concepts to explain or predict a given event, or give a broader understanding of the phenomenon of interest" (p. 189). At the same time, Imenda emphasizes that a framework forms the specific perspective which a researcher uses to explore, interpret or explain events or behaviours. Thus, it appears that a conceptual framework is a dual-nature notion: It may serve both as a tool for research and as a result of research. In other words, we may use a framework as a structured way to analyse and make sense of our data, but concurrently this process frames our perspective, resulting in a coherent system of ideas that we may also call *a framework*. Perhaps this dual nature is one of the reasons why this term is so often presented in an idiosyncratic and even confusing manner. Ravitch and Riggan (2016) relate to this confusion and offer to resolve it by looking at a conceptual framework as an *argument*, aimed to convince that the study matters and that the research means are appropriate and rigorous. However, Ravitch and Riggan themselves, in the preface to their book, refer to a conceptual framework as both a process and a product; therefore, the duality remains, and perhaps the overall lesson to be learned is that ambiguity might be an inherent property of the definition of frameworks. Nevertheless, to minimize such ambiguity, I suggest an analogy from calculus that may be useful here: even for a discontinuous and "messy" function, if we define a small enough neighbourhood, we might be able to find a "smooth" behaviour of the graph. Similarly, if we relate to a specific domain within the wide ocean of educational research, we are likely to be able to reach a reasonable, and, even more importantly, valuable definition of framework. In this chapter, the domain I refer to is the knowledge and practices of mathematics teachers (MTs) and mathematics teacher educators (MTEs). Therefore, for this particular domain, I define a framework as *a set of constructs that unpack various aspects of the knowledge and practices that MTs and MTEs develop within their work*. Within this "neighbourhood", frameworks are useful as organizers of our understanding about the complexities involved in MTs and MTEs work. Furthermore, adapting a framework from the MT level to the MTE level, can shed light on the differences and similarities between these two professions, as I shall elaborate later on.

1.3. Developing frameworks for the MTE level: a brief overview

As mentioned earlier, the need to refer to the MTE level in addition to the MT level emerged in recent years, as we witness the research on MTEs gradually accumulating and added to the extensive body of literature on MTs, already broadly developed. Much attention is currently given to various aspects of the MTE role, for instance, MTEs' preparation (e.g. Borko, Jacobs, Koellner & Swackhamer, 2015; Karsenty, 2016; Schüler & Rösken-Winter, 2018), MTEs' needed skills (e.g. Elliott et al., 2009; Lesseig et al., 2017; van Es, Tunney, Goldsmith & Seago, 2014), MTEs' evolving practices (e.g. Coles, 2019; Kuzle & Biehler, 2015; Prediger & Pöhler, 2019) and professional development processes of MTEs during their work period (e.g. Rösken-Winter, Schüler, Stahnke & Blömeke, 2015; Karsenty, 2018a). This increasing interest in MTEs necessitates the development of frameworks to understand and conceptualize the profession of MTEs and its relation to the profession of MTs.

But how may such frameworks be developed? Konuk (2018) describes four main approaches by which different researchers develop frameworks for conceptualizing the knowledge and/or practices of teacher educators (TEs) in general, and of MTEs in particular:

1   *The standards-based approach.* Researchers use standards that pertain to what TEs are required to do,[2] in order to identify categories of essential knowledge needed to fulfil these requirements and derive a framework for TEs' knowledge (e.g. Lunenberg, Dengerink & Korthagen, 2014, cited in Konuk, 2018). Konuk notes that although published standards are usually more skill-oriented than knowledge-oriented, they can nevertheless inform our understanding of what comprises TEs' knowledge. This approach, however, may result in frameworks that are less theoretical and more pragmatic in nature

2   *The inquiry-based approach.* Within this approach, conceptualization starts with raising inquiry questions regarding essential knowledge needed for TEs, including knowledge of theory and of research in the domain of teacher education. Although Konuk does not mention specific frameworks that exemplify this approach, I suggest that Even's (2008) formation of the *knowtice* construct, that signifies the essence of what MTEs need to learn and develop, falls under this category. I see the *knowtice* construct as a conceptual framework, although it is not originally defined as such, as it is clearly meant to unpack various aspects of the knowledge and practices of MTEs; thus it satisfies the definition given earlier in Section 1.2. Reading the detailed account that Even (2008) provides about the development of *knowtice*, it is apparent that it began with a series of fundamental inquiry questions

3   *The practice-oriented approach.* Here, constructing a framework begins with inspecting TEs' practices and identifying vital aspects of knowledge that emerge from them. For instance, John (2002) explored the work of six TEs with prospective teachers, in an attempt to capture the types of knowledge that underpin their role as teacher educators, resulting in a four-dimension framework

4   *Extending or revising existing teacher knowledge frameworks.* The principle of this approach is taking a framework that was originally created to conceptualize what the MT profession entails, in terms of knowledge, practices,

decision-making processes, etc., and adapting it to the MTE level by extending or revising its components. This approach appears to be gaining attention in recent years and has been employed by several researchers; thus in this chapter I focus on it, drawing on others' as well as on my own research. In what follows, I use the notion of "a double-level use of frameworks" to signify this approach

2. A double-level use of frameworks

The attempt to adapt or revise a framework that originally refers to MTs, in order to conceptualize MTEs' knowledge (and practices), involves some pre-assumptions regarding the relation between the MT and the MTE levels. I relate here to two of these assumptions.

The first assumption is that a connection exists between what a teacher needs to know and what a teacher educator needs to know; in other words, that certain components of the framework at the MT level correspond with components at the MTE level, otherwise there is little point in adapting the framework. However, such a connection may vary with regard to different types of MTEs (e.g. university-based mathematics educators, mathematicians, former mathematics teachers, mathematics education researchers; see Beswick & Chapman, 2015). For example, Jaworski (2008) described the connection between MTs' and MTEs' knowledge as a partial overlap, as portrayed in Figure 5.1.

Such partial overlap of knowledge seems to be typical when the MTE is not a former teacher but rather a university-based mathematics educator or a researcher. Beswick and Chapman (2012, cited in Jaworski & Huang, 2014) point to elements of MTs' knowledge that these MTEs do not necessarily need to know, such as the daily implementation of a particular curriculum (although they should know the theory and design behind that particular curriculum). However, the connection may be different when it comes to MTEs who are

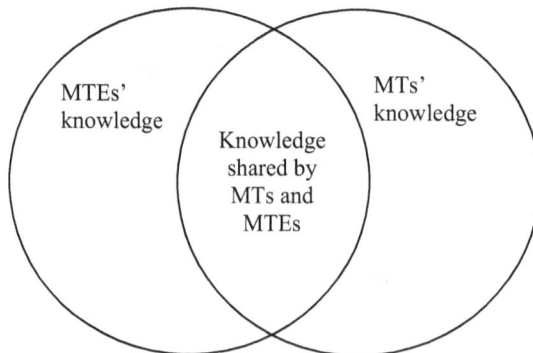

FIGURE 5.1 Interconnections between MTs' and MTEs' knowledge.

Source: Adapted from Jaworski (2008).

mathematics teachers that become educators of other teachers, mostly within in-service PD programmes. These MTEs are referred to in the literature in various terms, such as leaders or lead teachers (Borko, Koellner & Jacobs, 2011; Elliott et al., 2009; Lesseig et al., 2017; Zwetzschler, Rösike, Prediger & Barzel, 2016), mentor-teachers (Kuzle & Biehler, 2015), PD providers (Even, 2005), facilitators (Karsenty, 2016; Prediger et al., 2019; Schifter & Lester, 2002) and other terms (see Even, 2014). Herein I will use the term "facilitators" when referring to this particular group of MTEs. The correspondence between teachers' and facilitators' knowledge is typically not described as a partial overlap but as an expansion of knowledge. This relates to the already-established understanding that the transition from a teacher to a facilitator is far from being trivial (Arcavi, 2019; Even, 2005, 2008; Karsenty, 2016) and involves an identity change (Dinkelman, Margolis & Sikkenga, 2006; Murray & Male, 2005). As Murray and Male (2005) have put it, teachers who became TEs were "positioned as the *expert become novice*, in that they needed to acquire new knowledge and understanding" (p. 135, emphasis in original). Thus, facilitators are expected to develop knowledge at the PD level, while their expertise at the classroom level is valuable and is now nested within their new knowledge (Prediger et al., 2019). Perks and Prestage (2008) discuss this relationship, and based on their perspective, the relationship between MTs' and MTEs' knowledge can be portrayed visually as shown in Figure 5.2.

In sum, for adapting a framework from the MT level to the MTE level, researchers need to consider their theoretical assumption on the type of correspondence that exists between MTs' and MTEs' knowledge. As I shall exemplify in Section 3, this has important implications for the ways by which frameworks are adapted.

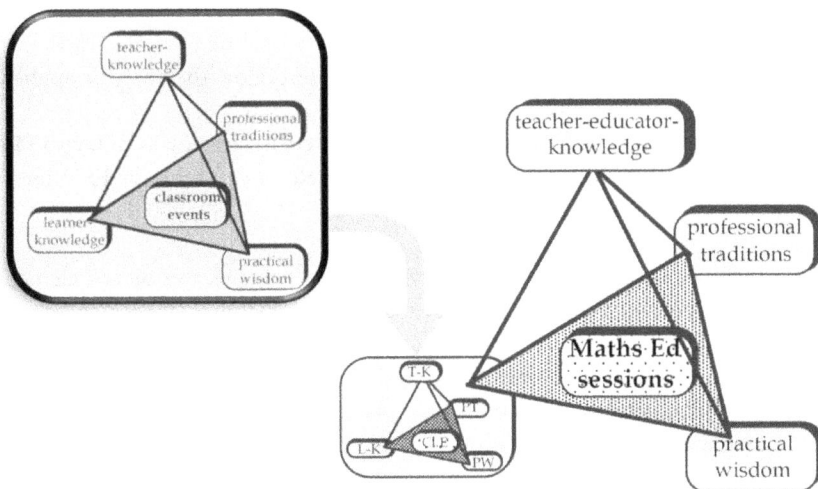

FIGURE 5.2 MTs' knowledge nested in MTEs' knowledge.

Source: Adapted from Perks and Prestage (2008).

The second assumption is that the adaptation of the framework is a worthwhile endeavour. This means that (a) the researcher ascribes value to the original framework at the MT level, i.e. views it as a powerful means to unpack and understand teacher knowledge and practices, and (b) the researcher believes that expanding, adapting or revising this framework will not only be powerful for understanding MTEs' knowledge and practices but will also be fruitful in shedding some light on the system as a whole, where a system is an integration of the various elements involved in the development of teacher learning, as indicated by Borko (2004). For example, Prediger et al. (2019) advocate that a multifaceted structure of teaching and learning, relating to different levels, is necessary for capturing the complexities involved in PD courses and for understanding specific design and research approaches and analysing their contribution. I suggest that working under such a system-based perspective is a pre-condition to the productive use of double-level frameworks.

3. Adapting frameworks from the classroom level to the PD level

At this point I wish to further "zoom in" into the process of creating a double-level framework, i.e. a process in which frameworks unpacking teachers' knowledge and practices are adapted or revised to include both the teacher and the facilitator levels. To do that, I borrow from the recent work of Prediger et al. (2019) that details three strategies for setting research agendas within a multi-level structure. These strategies are *lifting*, *nesting* and *unpacking*. While Prediger et al. (2019) relate these strategies to developing design principles, research practices and methods, I suggest they can be applied – and in fact, were applied in various cases, even if implicitly – to creating conceptual frameworks for the facilitator level. In particular, the lifting and nesting strategies, and sometimes their combination, are useful in the process of forming new frameworks. In the following, I describe these two strategies and exemplify how they may be applied.

The lifting strategy. Prediger et al. (2019) define this strategy as follows (TPD in this citation means Teacher Professional Development, whereas FPD means Facilitator Professional Development):

> lifting a design approach means that design principles or design elements developed for the classroom level are implicitly or explicitly transferred (and adapted) to the TPD level (or from the TPD to FPD level). Lifting a research practice means that certain types of research questions and/or methods from the classroom level are implicitly or explicitly transferred (and adapted) to the TPD level . . . and applied in an analogous way.
>
> *(pp. 412–413)*

Similarly, we can refer to lifting a framework as the idea of creating an analogy between an existing framework at the MT level and a new framework at the MTE level. As an example of such lifting, I use the case of the MKT and MKPD frameworks.

The MKT (Mathematical Knowledge for Teaching) framework was created by Ball, Themes and colleagues (Ball et al., 2008; Thames, Sleep, Bass & Ball, 2008) to unpack the knowledge needed by mathematics teachers, and it has become well known in the field of mathematics education and widely used in research on teacher learning. The MKT framework is comprised of six categories, divided into two main constructs: *subject matter knowledge*, which includes the categories Common Content Knowledge (CCK), Specialized Content Knowledge (SCK) and Horizon Content Knowledge (HCK); and *pedagogical content knowledge*, which includes the categories Knowledge of Content and Students (KCS), Knowledge of Content and Teaching (KCT) and Knowledge of Content and Curriculum (KCC). Borko, Koellner and Jacobs (2014) have suggested an analogous framework for unpacking knowledge required at the facilitator level, which they have named MKPD (Mathematical Knowledge for Professional Development). Thus, although not directly stated in Borko et al.'s (2014) work, it appears that the researchers have employed the lifting strategy. The MKPD framework encompasses three constructs: (a) *specialized content knowledge* (e.g. deep understanding of the mathematics that stands at the core of the PD and how to make it accessible to all PD participants), (b) *pedagogical content knowledge*, which, according to Borko et al. (2014), includes KCS and KCT but from a PD leader perspective (e.g. how to engage teachers in the interpretation of students' mathematical ideas and in productive analysis of instructional practices), and (c) *learning community knowledge* (e.g. how to establish group norms and foster active participation of teachers). The link between the MKT framework and the lifted framework of MKPD is portrayed visually in Figure 5.3.

Borko et al. (2014) explicitly refer to the relation between the two levels of knowledge, that of the teacher and that of the facilitator, asserting that the latter needs to

> go beyond and look different than the knowledge that a typical mathematics classroom teacher holds. Because PD leaders are expected to promote the development of teachers' knowledge in these domains, they must hold a deeper and more sophisticated knowledge of mathematics than their colleagues, just as teachers must hold a deeper and more sophisticated knowledge than their students.
>
> *(Borko et al., 2014, p. 165)*

Interestingly, although using an existing framework as a basis for adaptation, Borko and her colleagues testify that they conjectured about the nature of MKPD and the specifics of what its various components may entail, based on their analyses of teacher leaders' actual facilitation practices. This process can be described, in Konuk's (2018) terminology mentioned earlier, as employing a *practice-oriented approach*, which exemplifies that obtaining a double-level framework through adaptation can be carried out via one of the other three approaches indicated in Konuk (2018). In other words, Konuk's four categories are not mutually exclusive.

FACILITATOR	FACILITATOR	FACILITATOR
SUBJECT MATTER KNOWLEDGE	PEDAGOGICAL CONTENT KNOWLEDGE	LEARNING COMMUNITY KNOWLEDGE
Specialized Content Knowledge (SCK) (e.g., understanding the mathematics and making it accessible to PD participants)	Knowledge of Content and Students (KCS) Knowledge of Content and Teaching (KCT) (e.g., engaging PD participants in the interpretation of students' mathematical ideas and in productive analysis of instructional practices)	(e.g., establishing group norms and fostering active participation of PD participants)

MKPD: Facilitator level

SUBJECT MATTER KNOWLEDGE	PEDAGOGICAL CONTENT KNOWLEDGE
Common Content Knowledge (CCK) Specialized Content Knowledge (SCK) Horizon Content Knowledge	Knowledge of Content and Students (KCS) Knowledge of Content and Curriculum (KCC) Knowledge of Content and Teaching (KCT)

MKT: Teacher level

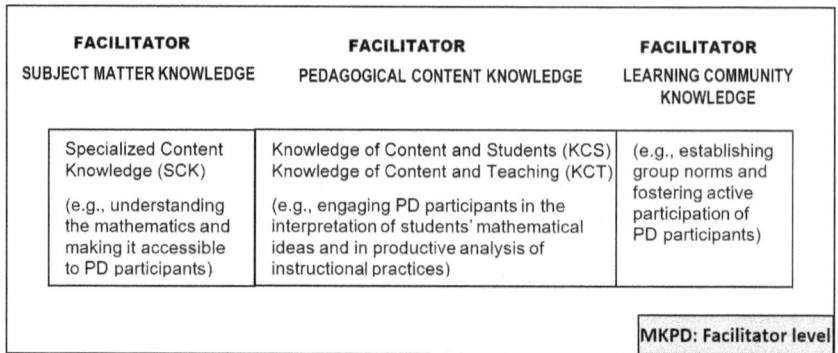

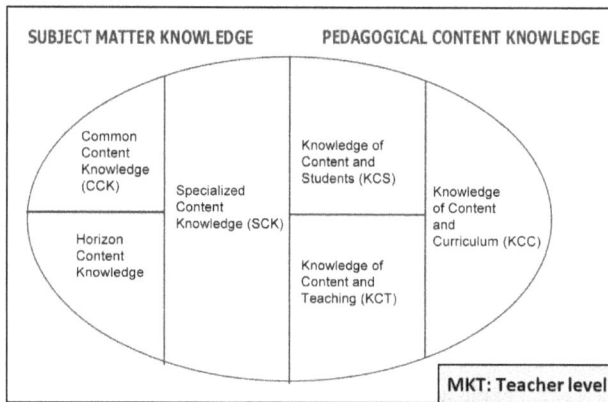

FIGURE 5.3 The MKT framework lifted into the MKPD framework.

Source: Lower box taken from Ball et al. (2008, p. 403); upper box based on Borko et al. (2014).

The nesting strategy. In general, nesting relates to the property of each set at a certain level being contained in the set at the next level. According to Prediger et al. (2019), the strategy of nesting means using a self-similar structure in different levels, so that the content, design principles or practices of one level are included as a component in the next level. Borrowing this strategy to the forming of double-level frameworks, nesting a framework that was originally created for the MT level within the MTE level, results in a complex structure where the elements that comprise the knowledge and practices of facilitators usually include, as a subset, the elements that comprise the knowledge and practices of teachers.

Searching the literature for examples of using the nesting strategy for conceptualizing the relations between the MT and the MTE levels reveals that mostly nesting has been used for expanding *models*, and not frameworks as they are defined here. Three examples of nested models are *the teaching triad* (Zaslavsky & Leikin, 2004), *the instructional triangle* (Ball, 2012) and the *nested domains of*

professional development programmes (Luft & Hewson, 2014), presented in Figures
5.4, 5.5 and 5.6 respectively.

As can be seen, all these examples present structural connections between the
MT and the MTE levels; however, they do not refer to specific components of the
knowledge and practices of MTs and MTEs, thus I classify them as models rather
than frameworks. Perks and Prestage's (2008) suggested structure of how MTs'
knowledge is nested in MTEs' knowledge (see Figure 5.2) is another example of
a model, though it is more detailed in regard to what MTEs knowledge includes.

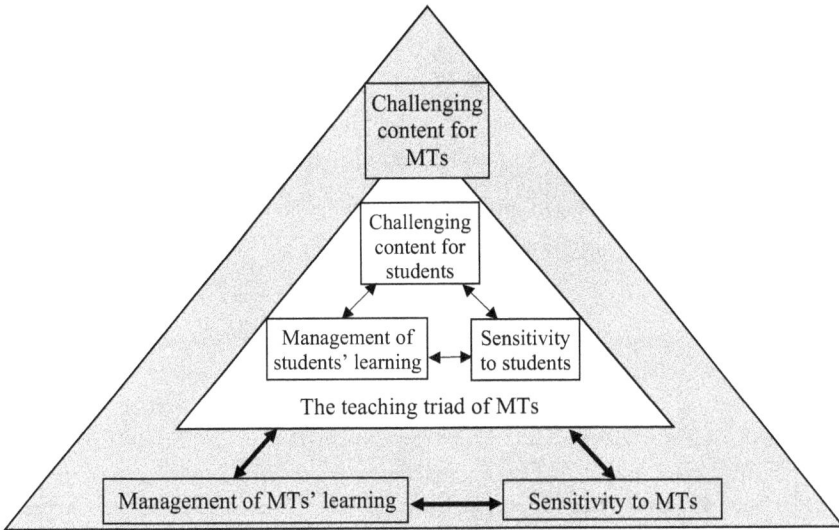

FIGURE 5.4   The teaching triad of MTs nested in the teaching triad of MTEs.

Source: Adapted from Zaslavsky and Leikin (2004).

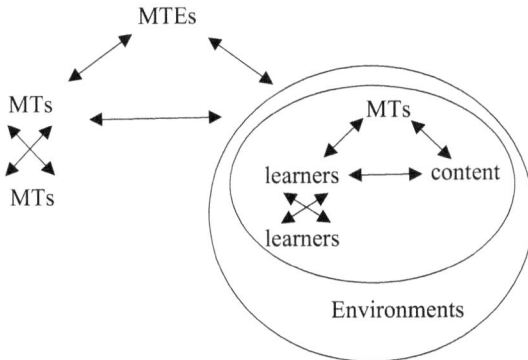

FIGURE 5.5   The instructional triangle of MTs nested in the instructional triangle of
MTEs.

Source: Adapted from Ball (2012).

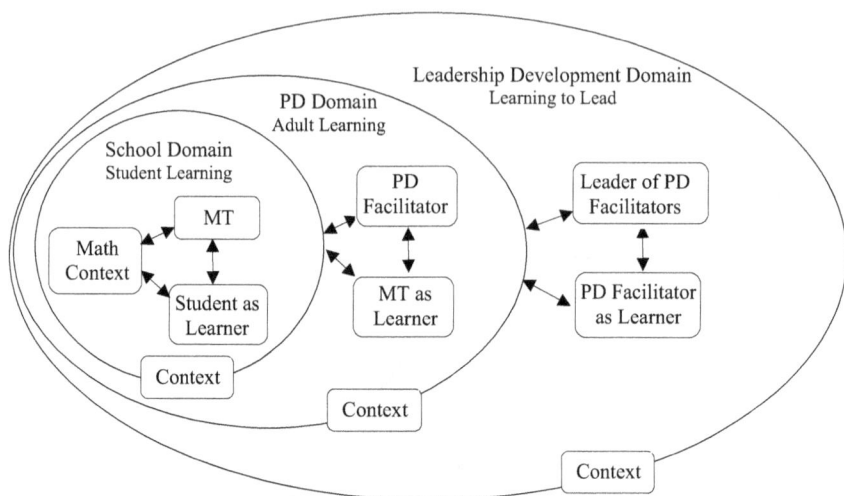

FIGURE 5.6 Nested domains in professional development.

Source: Adapted from Luft and Hewson (2014).

A unique example of a double-level framework that employs the nesting strategy (although not pertaining to PD facilitators but to educators of prospective teachers), is the Mathematical Knowledge for Teaching Teachers (MKTT), presented by Zopf (2010). The MKT framework (Ball et al., 2008) is nested in MKTT in the sense that MTEs' knowledge *includes* MKT; but more than that, MTEs need to possess the type of knowledge that "fosters teachers' decompression and unpacking of mathematical knowledge to develop it into the mathematical knowledge used for teaching" (Zopf, 2010, p. 194). Such MTE knowledge appears to entail the following:

- *Knowledge on how to unpack MKT for teachers*: this is described by Zopf as a fine-grained knowledge of interpretations and representations, coupled with knowledge of how to select examples for teachers that manifest and reveal these interpretations and representations.
- *A connected mathematical knowledge*: knowledge within and across domains that supports selection of details while working with teachers and threading these across mathematical ideas (for interesting examples of the use that MTEs make of such connected knowledge, see Zopf, 2010).
- *Knowledge on how to develop a precise mathematical language*: this requires knowing how to

> listen to, hear, and respond to teachers' mathematical talk; think through teachers' mathematical talk, discern what was correct and develop that language as well as discern what was incorrect and modify that into mathematically correct language; create an in-the-moment

example to make visible an error . . . foster mathematical language usable within and across mathematical domains.

(Zopf, 2010, p. 187)

- *Knowledge of the epistemology of mathematics*: this refers to robust knowledge about mathematical structures such as definitions, properties, theorems, etc., and knowledge about mathematical methods such as proof, explanations, justifications, etc., and how these are used for mathematical work.

MKTT is a framework intended to conceptualize the broad knowledge that MTEs use within their work. As in the case of MKPD described earlier, MKTT is an adaptation of an existing framework, originally constructed for the teacher level (MKT), employing a *practice-oriented approach* (Konuk, 2018), i.e. based on a careful analysis of MTEs observed practices. In the following section, I present in detail an example of adapting an existing framework and lifting it from the classroom level to the PD level, using an *inquiry-based approach*.

4. From lenses to meta-lenses: a case of adapting a framework from the teacher level to the facilitator level

In 2012, Abraham Arcavi and I began a research-based PD programme for secondary mathematics teachers in Israel. The aim of this endeavour was to enhance teachers' reflection on their practice and their MKT, under the working assumption that such enhancement will support mindful and well-informed decision-making by teachers while preparing and teaching lessons. The project, named VIDEO-LM (Viewing, Investigating and Discussing Environments of Learning Mathematics), was designed as a video-based PD, where videotaped mathematics lessons serve as learning objects and sources for discussions with teachers. Within the design phase, we realized we need a conceptual framework to be used for directing teachers' attention to various aspects of the lessons they observe. We asked ourselves, as designers, which elements will be necessary for unpacking mathematics teachers' knowledge and practices as they may be manifested in a lesson. Thus, our framework, which I detail in this section, was theory-driven; to use Konuk's (2018) terminology, it was formed through an inquiry-based approach.

We named the framework we created "the Six-Lens Framework", or in short, SLF. A detailed account of SLF can be found in several publications (e.g. Karsenty, 2017, 2018b; Karsenty & Arcavi, 2017); therefore, in this chapter I focus on how SLF was developed for the teacher level and later on adapted for the facilitator level.

SLF draws on Schoenfeld's (1998, 2010) *teaching in context* theory. Schoenfeld argues that teaching is goal-oriented; teachers strive to achieve various types of goals and are constantly modifying and changing their goals in correspondence with classroom realities. Schoenfeld also asserts that teachers have a body of knowledge resources they can call upon, for both expected and unexpected situations, and that teachers, like everyone else, have a set of orientations, i.e. predispositions and beliefs about mathematics, about students and about teaching. According to the teaching in context theory, the combination and inter-relation of goals, resources and orientations monitor teachers' decision-making processes and shape their choice of actions. Thus, it makes much sense that a PD programme aiming to enhance teacher reflection on practice should meaningfully involve explicit awareness to this triad. In their initial experimentation with video-based discussions that centralize these ideas, Arcavi and Schoenfeld (2008) have used several analytical components to direct mathematics teachers' reflection while watching videotaped lessons of unknown peers. Based on these experiments, we refined and extended the components into the conceptual framework of SLF. This framework consists of six "viewing lenses": (1) mathematical and meta-mathematical ideas around the lesson's topic; (2) explicit and implicit goals that may be ascribed to the teacher; (3) the tasks selected by the teacher and their enactment in class; (4) the nature of the teacher-student interactions; (5) teacher dilemmas and decision-making processes and (6) beliefs about mathematics, its learning and its teaching as inferable from the teacher's actions and reactions. Figure 5.7 presents these lenses and what each one of them is meant to unpack.

SLF is therefore a conceptual framework that organized our understanding as researchers as well as the understanding of teachers, about the complexities of teacher knowledge and practices. Moreover, our research around the VIDEO-LM project has shown that using SLF as a guiding framework supports the development of a reflective language among PD participants (Arcavi & Karsenty, 2018; Karsenty, 2017, 2018b; Karsenty & Arcavi, 2017; Schwarts & Karsenty, 2018, 2019). The following citation, from a teacher's end-of-course feedback, demonstrates this:

> These are really tools that now I use to look at lessons, and also when I plan lessons . . . everything suddenly has names. . . . There are many kinds of spectacles that now became natural to me.
>
> *(teacher citation taken from Karsenty, 2018b, p. 285)*

Two key phrases can be identified in this citation: "everything has names", and "spectacles that now became natural"; both are aligned with the definition of a framework as suggested in the beginning of this chapter, i.e. that a conceptual framework is a set of constructs that are meant to structure and unpack aspects of mathematics teachers' knowledge and practices.

During the first years of VIDEO-LM, PD courses were facilitated by the project team members. However, a growing demand for these courses resulted in upscaling the project to enable its availability to more teachers across the country, as of 2015. This necessitated the design and implementation of a system for preparing and supporting lead teachers to become VIDEO-LM facilitators (Karsenty, 2016). We have conducted two cohorts of preparation courses for qualifying new skilled facilitators. While designing the facilitator course, we

Lenses for the teacher level	What the lenses unpack
Mathematical and meta-mathematical ideas	The space of relevant ideas and concepts that underlie the topic of a mathematics lesson; meta-mathematical ideas (e.g., one counter example is sufficient to refute a conjecture) that are employed within the lesson.
Explicit and implicit teacher goals	Possible goals that may be attributed to the teacher, on the basis of actions or decisions observed in the lesson, as well as pros and cons of preferring certain goals over others.
Classroom tasks and activities	Features of the tasks and activities and how they are enacted in the lesson, including whether and when this process develops differently than expected ("a posteriori task analysis").
Teacher-student interactions	How the teacher poses further questions to those included in of the task; listens to (or ignores) comments or difficulties raised by students; manages discussions; delegates responsibilities in the process of knowledge generation.
Teacher dilemmas and decision-making	Teacher decisions prior to and during the lesson; situations of dilemma (i.e., when there is no evident optimal course of action) that the teacher seems to be facing during the lesson, and possible pathways that can be offered to resolve these dilemmas, while considering consequent tradeoffs.
Teacher beliefs about mathematics teaching, how students learn and the teacher's role	Orientations, beliefs and values that may be attributed to the teacher; implicit messages that may be conveyed to students through the teacher's communications and actions.

FIGURE 5.7 The Six-Lens Framework.

found ourselves again in need of a framework to conceptualize the knowledge and practices of VIDEO-LM facilitators. As with the work at the teacher level, we employed an inquiry-based approach, asking ourselves which elements will be necessary for identifying such knowledge and practices. This question was pursued not only to satisfy a theoretical interest; we had to teach prospective facilitators who, at this initial stage, needed to learn from the experiences of others. Therefore, it was our responsibility to supply them with vivid facilitation cases to examine and to assist them with developing the language and means to analyse these cases. We went through a process of comparing the knowledge and practices of mathematics teachers that constitute the SLF components, to the knowledge and practices needed for facilitators whose job is to use SLF with teachers. The result of this process was the creation of the *Meta-Lenses Framework* (MLF), portrayed in Figure 5.8.

Meta-lenses for the facilitator level	What the lenses unpack
The PD agenda, ideas and norms	What ideas that stand at the core of the project's agenda appear in the PD session. For example: the kind of reflection that is supported; the use made of the six lenses; the degree to which non-judgemental norms of discussion are followed; the video representations used.
Explicit and implicit facilitator goals	Possible goals that may be attributed to the facilitator, on the basis of actions or decisions observed in the PD session, as well as pros and cons of preferring certain goals over others.
PD tasks and activities	Features of the tasks and activities and how they are enacted in the PD session, including whether and when this process develops differently than expected.
Facilitator-teacher interactions	How the facilitator poses questions and manages the discussion; what facilitator moves are being employed; how the facilitator listens to (or ignores) comments raised by teachers and handles challenging situations such as norm violation; the degree to which the facilitator supports teacher collaboration.
Facilitator dilemmas and decision-making	Facilitator decisions prior to and during the PD session; situations of dilemma that the facilitator seems to be facing during the session, and possible pathways that can be offered to resolve these dilemmas, while considering consequent tradeoffs.
Facilitator beliefs about mathematics teaching, how teachers learn and the facilitator's role	Orientations, beliefs and values that may be attributed to the facilitator; implicit messages that may be conveyed to teachers through the facilitator's communications and actions.

FIGURE 5.8 The Meta-Lenses Framework.

MLF is an example of adapting an existing framework by lifting it from the class-room level to the PD level. As can be seen, there are differences and similarities between SLF and MLF. The major difference lies within the first lens, which relates to the subject matter at the core of learning. While the teacher is responsible for enhancing students' learning of mathematics, and thus the subject matter at the MT level clearly pertains to mathematical and meta-mathematical ideas (lens #1 in SLF), the case is different for the VIDEO-LM facilitator. Teachers are not enrolled in VIDEO-LM courses for the purpose of studying mathematics per se (this does not mean that they do not gain new mathematical knowledge through these courses; on the contrary, our research shows that as a result of participating in VIDEO-LM PDs, teachers enrich and crystalize their mathematical knowledge, which is in line with one of the project's aims. See Arcavi & Karsenty, 2018; Karsenty, Arcavi & Nurick, 2015). Teachers enrol in these PD courses in order to learn how to improve their practices by engaging in reflection and peer discussion. Therefore, the facili-tators are responsible for an entirely different domain of learning. They need to involve teachers in reflection, in using the SLF language, in developing norms for watching and discussing videotaped lessons and in maintaining productive discus-sions. Thus, the subject matter at the PD level is the ideas and norms that constitute the project's agenda (lens #1 in MLF). Schwarts (2020), reporting on a case of transition from a mathematics teacher to a PD facilitator, shows that such change in roles involves a growing awareness to the different epistemic status of these two subject matters, i.e. mathematics as a rigorous domain in which explicit rules are applied for determining the validity of claims, as opposed to PD ideas and norms which are much more flexible, subjective and context-dependent, and for which the dichotomy of "right" or "wrong" statements does not apply.

In a sense, the framework of SLF can be seen as nested in MLF, since all six lenses of the SLF are part of the subject matter for teacher learning within the PD, which constitutes the first lens of MLF. Thus, the double-level framework in this case employs both the lifting and nesting strategies, as portrayed in Figure 5.9. Moreover, it reflects a system-based perspective, discussed earlier in Section 2: the double level SLF + MLF framework can be valuable for understanding the complex interrelations of core elements involved in a PD system, i.e. teachers, facilitators, the programme's design and the context in which the PD operates (Borko, 2004). The following example demonstrates this affordance.

During 2015–16 we conducted a preparation course for high school teachers who wanted to become VIDEO-LM facilitators. In one of the course sessions, participants observed a classroom video in which the teacher dedicated most of the lesson time to a geometry task with multiple solution strategies. Then, the prospective facilitators watched a video of an experienced facilitator holding a PD session around this lesson, and encountering a challenge when teachers claimed that discussing several solution strategies with students is something they cannot afford to do. At this point, the video of the session was paused, and the prospective facilitators conducted a simulation centred on "What would you have done in this situation as the PD facilitator?" One participant, Olivia,[3] acted

Lenses for the teacher level		Meta-lenses for the facilitator level
Mathematical and meta-mathematical ideas		The PD agenda, ideas and norms
Explicit and implicit teacher goals		Explicit and implicit facilitator goals
Classroom tasks and activities		PD tasks and activities
Teacher-student interactions		Facilitator-teacher interactions
Teacher dilemmas and decision-making		Facilitator dilemmas and decision-making
Teacher beliefs about mathematics teaching, how students learn and the teacher's role		Facilitator beliefs about mathematics teaching, how teachers learn and the facilitator's role

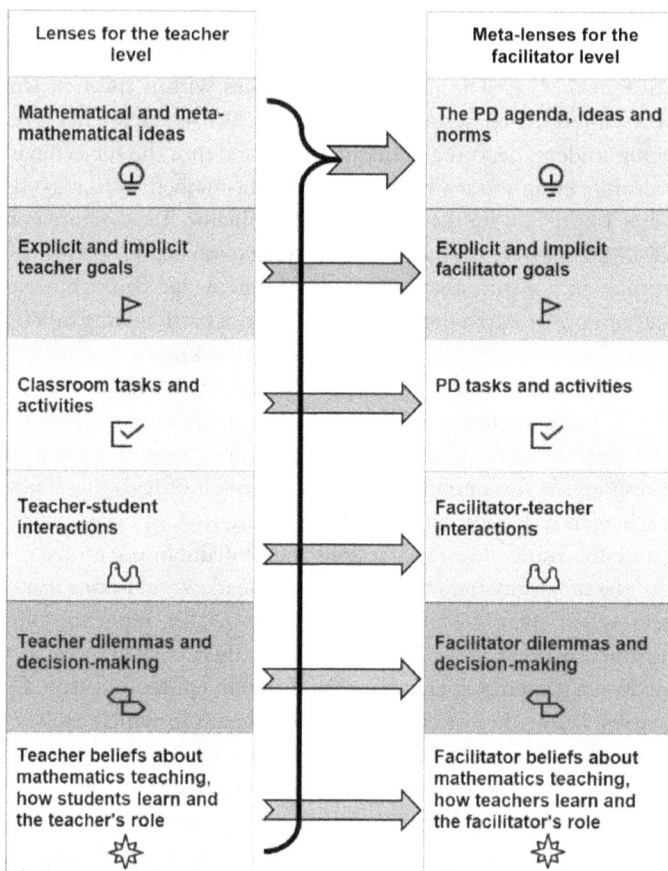

FIGURE 5.9 SLF and MLF: A double level framework.

as the facilitator, while other participants played the role of teachers. Olivia spent a considerable amount of time trying to offer pragmatic resolutions to the frustrating problem of time constraints, brought up by those who acted as teachers. When analyzing the simulation, the following discussion evolved between two prospective facilitators, Rita and Rebecca:

RITA: As a facilitator, I think I keep trying to solve frustrations people have, like, I want to have an answer for them, to save the day. And I suddenly realized that as a PD participant, I completely block myself if a facilitator tries to come and resolve my situation. And this is really, like, extreme how much I want to solve as a facilitator, and how unhelpful it is for me as a PD participant. . . . And even though Olivia did try to give the floor to the participants, it felt to me like it could have been done to a far greater extent . . . I mean . . . in any case, instead of saying what I think, [it's better] to ask "Is there someone who would have done this, or would have done that?" – to really open it up, and I suddenly realized how much I . . . I really do it poorly.

REBECCA: But on the other hand it's like when we're going to a concert, and the singer, you come to hear the singer, half a song he puts the microphone for the audience to sing, and this is also bothering. It's annoying.

RITA: But is a PD really a concert?

REBECCA: This is part of the issue . . . we joined the VIDEO-LM project, and whoever comes to these PD courses comes for several reasons, for one thing to learn and advance, and to see someone that, it's true he is like us, but he has . . . he comes from a level that is a little higher. . . . We just need to find, like we always say, the right balances, between opening it and not opening it, and even then you can't satisfy a hundred percent of the people.

Using the double level SLF + MLF framework to analyse this exchange can shed light on the complex system of teacher-facilitator relations. Through the double lens of *beliefs*, it appears that Rita finds the role of teacher and the role of facilitator to be occasionally in conflict; whereas a facilitator strives to "solve frustrations people have", teachers might view such an attempt as "unhelpful". Rita's inclination to prefer the teacher stance in this case has significant implications on how she perceives the *facilitator-teacher interactions*, as manifested in her suggestion to encourage teachers to talk rather than say what the facilitator herself thinks. The implicit *facilitator goal* here seems to be that teachers will use their own collective wisdom to resolve difficulties involved in their practice. This perspective is challenged by Rebecca, who links between teachers' aim when entering a PD, i.e. to learn from someone that "comes from a level that is a little higher", and the desired *facilitator's decision* to find "the right balances" between her own input and that of the teachers. Interestingly, while Rita speaks of the discrepancy between being a teacher and being a facilitator in terms such as "extreme", Rebecca maintains that teachers view the facilitator as someone that is basically "like us", which implies that a facilitator is expected to keep her identity as a teacher along with her new identity as a facilitator. Although not evident from this vignette, one can hypothesize that the differences between the two perspectives expressed by Rita and Rebecca would have an impact on how they may, as future facilitators, enact the project's *agenda and ideas* and handle facilitator *dilemmas*. Such hypotheses are currently being investigated in the course of an ongoing research we conduct on professionalization processes of facilitators, that will hopefully yield further insights on the complexities of PD systems.

5. Concluding words

This chapter focused on the notion of conceptual frameworks, in the specific context of the knowledge and practices of mathematics teachers and mathematics teacher educators. The idea of a double-level framework was examined from several aspects: how such frameworks are created, what purposes can they serve and what is their role in shedding light on phenomena observed when looking at the system of students–teachers–teacher educators as a whole. For example, questions such as "what happens when teachers become learners in PD settings" or "what happens when teachers become facilitators responsible for the learning of other teachers" are intriguing questions that are gaining more and more research attention. For understanding how such questions can be dealt with, the use of double-level frameworks may prove productive.

At the beginning of this chapter I conjectured that the apogee of the current trend – studying the MTE profession from various angles – is still ahead of us. If this is true, I expect that within the next decade we will witness more and more uses of double-level frameworks (for instance, the MTSK, Mathematics Teacher's Specialized Knowledge (Carrillo et al., 2018) may be productive as a basis for developing MTESK, Mathematics Teacher Educator's Specialized Knowledge). In what ways would such uses advance the community's knowledge and ability to develop and support the work of MTEs, is a question yet to be answered.

Notes

1 See also chapters by Carrillo and by de Gamboa et al., included in this volume.
2 For instance, the Association of Teacher Educators (ATE) in the United States has published nine standards for accomplished TEs. These standards were developed in 1992 and revised in 2008 (see Association of Teacher Educators, 2019). Similar standards appear in the Dutch context (Lunenberg, Dengerink & Korthagen, 2014, cited in Konuk, 2018).
3 Pseudonym, as are all participants' names.

References

Arcavi, A. (2019). From tools to resources in the professional development of mathematics teachers: General perspectives and crosscutting issues. In S. Llinares (Ed.), *The international handbook of mathematics teacher education* (2nd ed., pp. 421–440). Vol. 2: Tools and Processes in Mathematics Teacher Education. Leiden, The Netherlands: Brill Sense.

Arcavi, A., & Karsenty, R. (2018). Enhancing mathematics teachers' reflection and knowledge through peer-discussions of videotaped lessons: A pioneer program in Israel. In N. Movshovitz-Hadar (Ed.), *K-12 mathematics education in Israel: Issues and challenges* (Chapter 33, pp. 303–310). Series on Mathematics Education Vol. 13. Singapore: World Scientific.

Arcavi, A., & Schoenfeld, A. H. (2008). Using the unfamiliar to problematize the familiar: The case of mathematics teacher in-service education. *Canadian Journal of Science, Mathematics, and Technology Education, 8*(3), 280–295.

Association of Teacher Educators. (2019). *Standards for teacher educators.* Retrieved October 2019, from, https://ate1.org/standards-for-teacher-educators

Ball, D. L. (2012). Afterword: Using and designing resources for practice. In G. Gueudet, B. Pepin, & L. Trouche (Eds.), *From text to "lived" resources, mathematics teacher education* 7 (pp. 349–359). Dordrecht: Springer.

Ball, D. L., Thames, M. H., & Phelps, G. (2008). Content knowledge for teaching: What makes it special. *Journal of Teacher Education, 59*(5), 389–407.

Beswick, K., & Chapman, O. (2012). Mathematics teacher educators' knowledge for teaching. Paper presented at *the 12th International Congress on Mathematics Education*, Coex, Seoul, Korea.

Beswick, K., & Chapman, O. (2015). Mathematics teacher educators' knowledge for teaching. In S. G. Cho (Ed.), *Proceedings of the 12th International Congress on Mathematical Education* (pp. 629–632). Heidelberg, Germany: Springer.

Borko, H. (2004). Professional development and teacher learning: Mapping the terrain. *Educational Researcher, 33*(8), 3–15.

Borko, H., Jacobs, J., Koellner, K., & Swackhamer, L. (2015). *Mathematics professional development: Improving teaching using the problem-solving cycle and leadership preparation models*. New York: Teachers College Press.

Borko, H., Koellner, K., & Jacobs, J. (2011). Meeting the challenges of scale: The importance of preparing professional development leaders. *Teachers College Record*. Retrieved from www.tcrecord.org

Borko, H., Koellner, K., & Jacobs, J. (2014). Examining novice teacher leaders' facilitation of mathematics professional development. *The Journal of Mathematical Behavior, 33*, 149–167.

Carrillo, J., Climent, N., Montes, M., Contreras, L. C., Flores-Medrano, E., Escudero-Ávila, D. . . . Muñoz-Catalán, M. C. (2018). The Mathematics Teacher's Specialized Knowledge (MTSK) model. *Research in Mathematics Education, 20*(3), 236–253.

Coles, A. (2019). Facilitating the use of video with teachers of mathematics: Learning from staying with the detail. *International Journal of STEM Education, 6*(1), 1–12. doi:10.1186/s40594-018-0155-y

Da Ponte, J. P., & Chapman, O. (2006). Mathematics teachers' knowledge and practices. In A. Gutierrez & P. Boero (Eds.), *Handbook of research on the psychology of mathematics education* (pp. 461–494). Rotterdam, The Netherlands: Sense Publishers.

Dinkelman, T., Margolis, J., & Sikkenga, K. (2006). From teacher to teacher educator: Reframing knowledge in practice. *Studying Teacher Education, 2*(2), 119–136.

Elliott, R., Kazemi, E., Lesseig, K., Mumme, J., Carroll, C., & Kelley-Petersen, M. (2009). Conceptualizing the work of leading mathematical tasks in professional development. *Journal of Teacher Education, 60*(4), 364–379.

Even, R. (2005). Integrating knowledge and practice at MANOR in the development of providers of professional development for teachers. *Journal of Mathematics Teacher Education, 8*(4), 343–357.

Even, R. (2008). Facing the challenge of educating educators to work with practicing mathematics teachers. In B. Jaworski & T. Wood (Eds.), *International handbook of mathematics teacher education: The mathematics teacher educator as a developing professional* (Vol. 4, pp. 57–73). Rotterdam, The Netherlands: Sense Publishers.

Even, R. (2014). Challenges associated with the professional development of didacticians. *ZDM Mathematics Education, 46*, 329–333.

Even, R., & Krainer, K. (2014). Education of mathematics teacher educators. In S. Lerman (Ed.), *Encyclopedia of mathematics education* (pp. 202–204). Heidelberg: Springer.

Goos, M. (2009). Investigating the professional learning and development of mathematics teacher educators: A theoretical discussion and research agenda. In R. Hunter, B. Bicknell, & T. Burgess (Eds.), *Crossing divides: Proceedings of the 32nd Annual Conference of the Mathematics Education Research Group of Australasia* (Vol. 1, pp. 209–216). Palmerston North, NZ: MERGA.

Goos, M., & Beswick, K. (Eds.). (In press). *The learning and development of mathematics teacher educators: International perspectives and challenges*. New York: Springer.

Imenda, S. (2014). Is there a conceptual difference between theoretical and conceptual frameworks? *Journal of Social Sciences, 38*(2), 185–195.

Jaworski, B. (2008). Development of the mathematics teacher educator and its relation to teaching development. In B. Jaworski & T. Wood (Eds.), *International handbook of mathematics teacher education: The mathematics teacher educator as a developing professional* (Vol. 4, pp. 335–361). Rotterdam, The Netherlands: Sense Publishers.

Jaworski, B., & Huang, R. (2014). Teachers and didacticians: Key stakeholders in the processes of developing mathematics teaching. *ZDM Mathematics Education, 46*(2), 173–188.

Jaworski, B., & Wood, T. (Eds.). (2008). *International handbook of mathematics teacher education: The mathematics teacher educator as a developing professional* (Vol. 4). Rotterdam, The Netherlands: Sense Publishers.

John, P. D. (2002). The teacher educator's experience: Case studies of practical professional knowledge. *Teaching and Teacher Education, 18*(3), 323–341.

Karsenty, R. (2016, November). Preparing facilitators to conduct video-based professional development for mathematics teachers: Needs, experiences and challenges. Paper presented at *the 2nd International Conference on Educating the Educators*, Freiburg, Germany.

Karsenty, R. (2017). How do mathematics teachers learn from videotaped lessons of unknown peers? Exploring possible mechanisms that contribute to change in teachers' perspectives. In L. Gómez Chova, A. López Martínez, & I. Candel Torres (Eds.), *Proceedings of the Ninth Annual International Conference on Education and New Learning Technologies* (pp. 1718–1728). Barcelona, Spain: IATED Academy.

Karsenty, R. (2018a, June). Talking about observed practices: Enhancing novice facilitators' proficiency to steer video-based discussions with mathematics teachers. Paper presented at *EARLI SIG-11 Conference (Teaching and Teacher Education)*, University of Agder, Kristiansand, Norway.

Karsenty, R. (2018b). Professional development of mathematics teachers: Through the lens of the camera. In G. Kaiser, H. Forgasz, M. Graven, A. Kuzniak, E. Simmt, & B. Xu (Eds.), *Invited lectures from the 13th International Congress on Mathematical Education* (pp. 269–288). Hamburg: Springer.

Karsenty, R., & Arcavi, A. (2017). Mathematics, lenses and videotapes: A framework and a language for developing reflective practices of teaching. *Journal of Mathematics Teacher Education, 20*, 433–455.

Karsenty, R., Arcavi, A., & Nurick, Y. (2015). Video-based peer discussions as sources for knowledge growth of secondary teachers. In K. Krainer & N. Vondrová (Eds.), *Proceedings of the Ninth Congress of the European Society for Research in Mathematics Education* (pp. 2825–2832). Prague: ERME.

Konuk, N. (2018). *Mathematics teacher educators' roles, talks, and knowledge in collaborative planning practice: Opportunities for professional development* (Unpublished doctoral dissertation in curriculum and instruction). Pennsylvania State University. Retrieved from https://etda.libraries.psu.edu/catalog/15800nuk141

Kuzle, A., & Biehler, R. (2015). Examining mathematics mentor teachers' practices in professional development courses on teaching data analysis: Implications for mentor teachers' programs. *ZDM Mathematics Education, 47*(1), 39–51.

Lesseig, K., Elliott, R., Kazemi, E., Kelley-Petersen, M., Campbell, M., Mumme, J., & Carroll, C. (2017). Leader noticing of facilitation in videocases of mathematics professional development. *Journal of Mathematics Teacher Education, 20*(6), 591–619.

Luft, J. A., & Hewson, P. W. (2014). Research on teacher professional development in science. In S. K. Abell & N. G. Lederman (Eds.), *Handbook of research in science education* (Vol. 2, pp. 889–909). New York, NY: Routledge.

Lunenberg, M., Dengerink, J., & Korthagen, F. (2014). *The professional teacher educator: Roles, behavior, and professional development of teacher educators.* Rotterdam, The Netherlands: Sense Publishers.

Maxwell, J. A. (2005). Conceptual framework: What do you think is going on? In J. A. Maxwell (Ed.), *Qualitative research design: An interactive approach* (3rd ed., pp. 39–72). Thousand Oaks, CA: Sage Publications. Retrieved from www.sagepub.com/sites/default/files/upm-binaries/48274_ch_3.pdf

Murray, J., & Male, T. (2005). Becoming a teacher educator: Evidence from the field. *Teaching and Teacher Education, 21*(2), 125–142.

Perks, P., & Prestage, S. (2008). Tools for learning about teaching and learning. In B. Jaworski & T. Wood (Eds.), *International handbook of mathematics teacher education: The mathematics teacher educator as a developing professional* (Vol. 4, pp. 265–280). Rotterdam, The Netherlands: Sense Publishers.

Prediger, S., & Pöhler, B. (2019). Conducting PD discussions on language repertoires: A case on facilitators' practices. In M. Graven, H. Venkat, A. Essien, & P. Vale (Eds.), *Proceedings of the 43rd Conference of the International Group for the Psychology of Mathematics Education* (Vol. 3, pp. 241–248). Pretoria, South Africa: PME.

Prediger, S., Roesken-Winter, B., & Leuders, T. (2019). Which research can support PD facilitators? Strategies for content-related PD research in the three-tetrahedron model. *Journal of Mathematics Teacher Education, 22*, 407–425.

Ravitch, S. M., & Riggan, M. (2016). *Reason & rigor: How conceptual frameworks guide research.* Thousand Oaks, CA: Sage Publications.

Rösken-Winter, B., Schüler, S., Stahnke, R., & Blömeke, S. (2015). Effective CPD on a large scale: Examining the development of multipliers. *ZDM Mathematics Education, 47*(1), 13–25.

Schifter, D., & Lester, J. B. (2002). Active facilitation: What do facilitators need to know and how might they learn it? *The Journal of Mathematics and Science: Collaborative Explorations, 8*, 97–118.

Schoenfeld, A. H. (1998). Toward a theory of teaching-in-context. *Issues in Education, 4*(1), 1–94.

Schoenfeld, A. H. (2010). *How we think: A theory of goal-oriented decision making and its educational applications.* New York, NY: Routledge.

Schoenfeld, A. H. (2018). Video analyses for research and professional development: The Teaching for Robust Understanding (TRU) framework. *ZDM Mathematics Education, 50*(3), 491–506.

Schüler, S., & Rösken-Winter, B. (2018). Compiling video cases to support PD facilitators in noticing productive teacher learning. *International Journal of STEM Education, 5.* doi:10.1186/s40594-018-0147-y

Schwarts, G. (2020). Facilitating a collaborative professional development for the first time. In H. Borko & D. Potari (Eds.), *Teachers of Mathematics Working and Learning in Collaborative Groups, Proceedings of the 25th ICMI Study Conference* (pp. 540–547). Lisbon, Portugal: ICMI.

Schwarts, G., & Karsenty, R. (2018). A teacher's reflective process in a video-based professional development program. In E. Bergqvist, M. Österholm, C. Granberg, & L. Sumpter (Eds.), *Proceedings of the 42nd Conference of the International Group for the Psychology of Mathematics Education* (Vol. 4, pp. 123–130). Umeå, Sweden: PME.

Schwarts, G., & Karsenty, R. (2019). "Can this happen only in Japan?": Mathematics teachers reflect on a videotaped lesson in a cross-cultural context. *Journal of Mathematics Teacher Education.* https://doi.org/10.1007/s10857-019-09438-z

Thames, M. H., Sleep, L., Bass, H., & Ball, D. L. (2008). Mathematical knowledge for teaching (K-8): Empirical, theoretical, and practical foundations. Paper presented at *the 11th International Congress on Mathematics Education* (*TSG 27*), Montérrey, Mexico.

van Es, E. A., Tunney, J., Goldsmith, L. T., & Seago, N. (2014). A framework for the facilitation of teachers' analysis of video. *Journal of Teacher Education, 65*(4), 340–356.

Zaslavsky, O., & Leikin, R. (2004). Professional development of mathematics teacher educators: Growth through practice. *Journal of Mathematics Teacher Education, 7*(1), 5–32.

Zopf, D. (2010). *Mathematical knowledge for teaching teachers: The mathematical work of and knowledge entailed by teacher education* (Unpublished doctoral dissertation). University of Michigan, Ann Arbor.

Zwetzschler, L., Rösike, K.-A., Prediger, S., & Barzel, B. (2016). Professional development leaders' priorities of content and their views on participant-orientation. Paper presented in *TSG 50 at ICME 13*, Hamburg.

6

PARALLEL STORIES

Teachers and researchers searching for mathematics teachers' specialized knowledge

José Carrillo

1. Introduction

It is often the case that researchers present their work and all its paraphernalia (theoretical frameworks, analytical results, models and the rest) without shedding any light on the original impulse which set their studies in motion. Here, I'm not referring to the underlying theory or methodology chosen for carrying out the research, but to the circumstances which caused the research group to choose one area of focus over another, and even sometimes which particular paradigm to employ.

How does it come about that a particular researcher settles on a specific case and sees it through to the end of the investigative process? All research has its backstory, although when it comes to writing up our results into academic papers, we do not always do justice to the original contexts from which our work grew. We typically couch our aims in terms of the interests and areas of study prioritized within the field of mathematics education, but in reality these are more often contingent upon the trajectory of the research group, the priorities of national or international research councils, or the group of teachers with whom the research group collaborates.

In this chapter I would like to draw attention to this latter source. Evidently, it is not exclusive, as influences tend to be multiple and are often difficult to disentangle, but I would like to bring to the fore – without disesteeming other influences – the role that a group of teachers can have in directing research interests, as this important source is given scant attention in the literature.

2. The search for key factors

The search for key factors in understanding a teacher's classroom performance goes beyond the understanding itself to encompass the search for elements and tools of use to professional development and teacher training, where the objective is to provide teachers with the knowledge and skills they need to carry out their profession.

When I first became a researcher, my interests, like those of the research group at the University of Huelva, can be described as the search for the key factors for understanding mathematics teachers' work with their students. Our first approaches are documented in Carrillo and Contreras (1994) and Contreras, Carrillo and Guevara (1996), which presented research tools for analysing teachers' conceptions about mathematics, teaching and learning mathematics and the use of problem-solving in lessons. These tools used descriptors to differentiate the conceptions according to four teaching styles – traditional, technological, spontaneous and investigative – across six categories: methodology, subject significance, learning conception, student's role, teacher's role and assessment. A further set of descriptors, drawn from Ernest (1991), distinguished three categories of conceptions about mathematics – instrumentalist, Platonic and problem-solving or dynamic – in terms of the type of knowledge, the aims of the mathematics involved and the way it was developed.

We were convinced of the importance of teachers' conceptions for understanding lessons in teacher-focused research (e.g. Lubinski & Vacc, 1994). At the same time, I explored my own methodological position in favour of problem-solving at all educational levels with studies into how teachers performed as solvers of problems themselves (as opposed to their pupils). The result of this intersection of interests was the consideration of the role that awareness of one's own conceptions, allied to the teacher putting into practice problem-solving based on genuine problems, can play as a trigger for professional development (Carrillo, 1999).

It was also during this period that, in 1999, our long collaboration with a group of teachers began. The group, known as the PIC (from the Spanish "Proyecto de Investigación Colaborativa"), continues to flourish today, having gone from two primary teachers and two researchers in its original composition, to its current constitution of various teachers at all educational levels (two pre-school teachers, three primary teachers, three secondary teachers), a primary education trainee, a schools inspector, a master student, a doctoral student and three researchers. The dynamic of the PIC is characterized by the meetings every 2 weeks and the design, implementation, observation and analysis of lessons, apart from the discussion on our conceptions on mathematics teaching and learning in an equitable atmosphere. Key elements in the PIC are the commitments of their participants and the negotiation of goals (Goos, 2014).

The PIC brought about substantial change in our interests and in our approach to researching mathematics education. Teachers ceased to be the object of our study and became instead collaborators in our research (Feldman, 1993; Climent & Carrillo, 2002). Working alongside teachers in a context focused on their classroom experiences led us to become interested in delineating professional development at the same time as we sought to promote it. It was in this way that we adopted the mode of research we continue to employ today, one pertaining to the *developmental* paradigm (Jaworski, 2005).

Not all members of the group shared the same interests regarding the process of research and professional development, however. The primary teachers were reluctant to make the main focus of their reflection the subject of mathematics itself, as it touched on their insecurities. Thus it became necessary to find an alternative means of developing their professional interests in a way satisfactory to both parties and which would avoid tackling – at least directly – their knowledge of mathematics.

Clearly, teachers' knowledge of mathematics is instrumental to their teaching the subject, but it is also true that reflection about this knowledge can be channelled through other perspectives, such as consideration of pupils' learning processes when they take on the challenge of mathematics problems. It was precisely this scenario which triggered the formation of the PIC, when the original primary teachers sought out our support for their plan to implement a problem-solving approach in their lessons, an approach for which they felt they had little practical training. Thus the group was established, with the first few sessions devoted to sharing conceptions of teaching and learning mathematics and to reaching a consensus on what would be understood as a problem.

The subsequent lesson plans based on problem-solving were (and still are) an excellent vehicle for reflection (albeit indirectly) on the knowledge of each of the PIC participants. Nevertheless, as the knowledge this procedure revealed was inevitably somewhat diffuse, it became replaced as a marker of professional development by the teacher's capacity for reflection (Climent, 2005), with indicators scaled in terms of increasing complexity and dimensionality.

There can be no doubt that the professional development of mathematics teachers has long been a fruitful area of research (see, for example, the *Journal of Mathematics Teacher Education*, CERME proceedings or Hospesová, Carrillo & Santos, 2018), but in our case, putting aside the interest we share with a lot of researchers, the not inconsiderable weight of the practising teachers participating in the PIC – and their motivation to professionally develop and so enhance their pupils' learning – also needs to be taken into account. Indeed, the focus on reflection was very much prompted by their personal predispositions, being inclined towards a discursive approach to sessions and disinclined towards anything that might (they feared) expose their mathematical shortcomings.

The collaborative ethos of the PIC governed the way in which we selected the areas we would focus on and the manner in which we would approach them, and also, above all, the objectives. As researchers, we felt it was not our role to establish the criteria for defining success (such as fulfilling indicators on a scale of teaching styles; Carrillo & Contreras, 1994; Ponte & Chapman, 2006). It should be the PIC as a group that defined its aims and the means of achieving them.

In this respect the teachers needed to make the research project and the process of professional development their own. We did not conceive of any other way of going about things, and we could not appropriate this responsibility (Keffer,

Wood, Carr, Mattison & Lanier, 1998), as it represented the linchpin to guaranteeing a certain degree of sustainability in the enterprise (Zehetmeier, 2010). In this way the question of how to define what the PIC understood as good practice emerged. It was not simply a matter of pinpointing particular features but more a project directed at capturing, whenever classroom performance was analysed, those aspects that could be considered as exemplifying good classroom practice. Examples of what was agreed by teachers and researchers within the PIC include a focus on different kinds of mathematics within a predominantly problem-solving approach (which is in line with the teachers' original plan, as mentioned earlier), a variety of activities drawn as much as possible from real-life contexts, teaching strategies favouring exploration and classroom management promoting the participation of all pupils in the mathematical tasks and a sense of involvement in their own learning (Carrillo & Climent, 2011).

In 2009, after the PIC had been in operation for several years, the teacher members came to the conclusion that, despite being aware of having moved forward (they attested to feeling more in confident in class, to being more self-critical and to planning more effective problem-solving tasks), there remained an obstacle which impeded any further real development – their knowledge of mathematics. We researchers met this admission with surprise and gratitude. The teachers made it clear that they did not want to study mathematics for its own sake, but they were open to the idea of testing the limits of their knowledge in relation to the requirements of carrying out problem-solving tasks in class, and to the extent that they enhanced their pupils' learning experiences.

3. What knowledge do teachers talk about?

When we talk about teachers' knowledge, it is as well to specify what knowledge we are referring to. Shulman's (1986) components of professional knowledge provide us with a precise terminology for talking about the knowledge all teachers need and bring into play in the course of their work. Whilst recognizing the equal importance of all these components (such as general pedagogical knowledge), the teachers and researchers in the PIC agreed to focus only on those components which were concerned with mathematical content. This focus intersected with the notion of the mathematics teacher's specialized knowledge.

In this respect we diverged from Ball, Thames and Phelps's (2008) formulation of specialized knowledge, which locates such knowledge within a sub-domain of mathematical knowledge. In our view, specialization is intrinsic to the profession, that is, an item of knowledge is specialized to the extent that it both serves teaching and derives from mathematics (Scheiner, Montes, Godino, Carrillo & Pino-Fan, 2019). The model we developed to reflect this view recognized the influence of several previous contributions to the field. It retained the division of MKT (Mathematical Knowledge for Teaching) (Ball et al., 2008) into two domains (mathematical knowledge and pedagogical content knowledge) and added a third, concerned with beliefs (about mathematics itself for

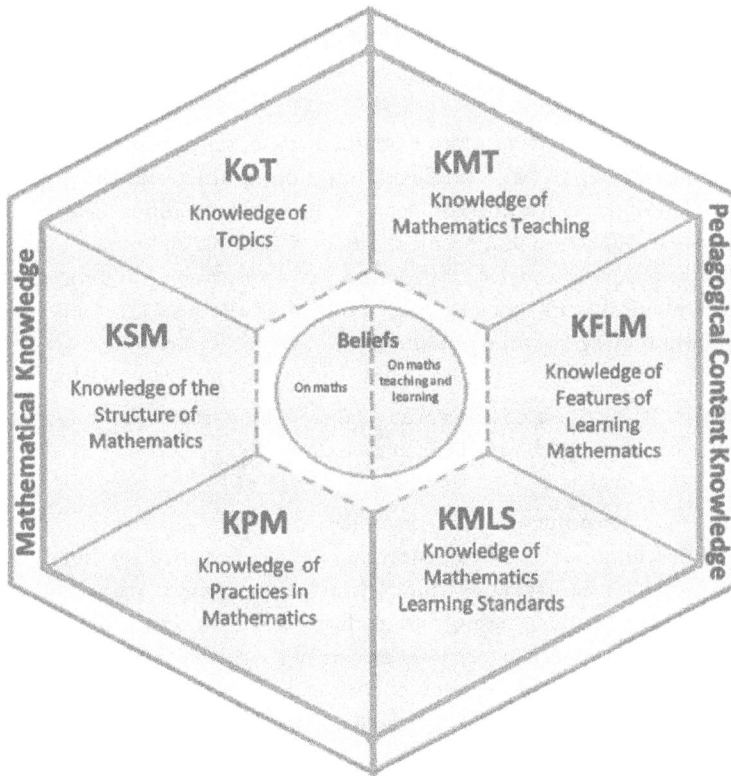

FIGURE 6.1 The MTSK model.

one part, and about the teaching and learning of the subject for the other). Also influential were Rowland, Turner, Thwaites and Huckstep's (2009) notion, within the Knowledge Quartet, of *contingency* – unexpected classroom situations bringing a challenge to teachers' knowledge – and Ma's (1999) knowledge packages, which we extended to all knowledge. In keeping with our ethos, the research group, coordinated at the University of Huelva, carried out various studies (some directly involving the PIC), which adopted a case-study design and drew on grounded theory (Charmaz, 2014) and a *top-down, bottom-up* approach (Grbich, 2013). The result was the Mathematics Teacher's Specialized Knowledge (MTSK) (Figure 6.1).

The MTSK model considers three subdomains within Mathematical Knowledge, each demarking the teacher's knowledge as follows:

- School topics (Knowledge of Topics – KoT): knowledge of procedures, definitions, properties (and their foundations), registers of representation, phenomenology and applications

- Interconnections between topics (including concepts and procedures) (Knowledge of the Structure of Mathematics – KSM): knowledge of connections involving increased complexity or simplification, and cross-curricular and "auxiliary" links
- Ways of doing and generating mathematics (Knowledge of Practices in Mathematics – KPM): knowledge of, among other aspects, hierarchical mechanisms and planning in relation to solving mathematical problems, modes of validation and proof, the role of symbols and use of formal language, processes involved in solving problems, particular procedures for mathematical work (for example, modelling), the necessary and sufficient conditions for generating definitions

For its part, Pedagogical Content Knowledge is composed of the following subdomains, according to the teacher's knowledge of:

- Strategies, techniques, tasks, examples, resources and theories associated with teaching mathematics (Knowledge of Mathematics Teaching – KMT)
- Pupils' strengths, weaknesses, interests and expectations in relation to content, and their modes of engaging with mathematical matters and theories of learning mathematics (Knowledge of Features of Learning Mathematics – KFLM)
- Learning expectations for each educational stage, the conceptual or procedural demands of each stage and the sequential order of topics (Knowledge of Mathematics Learning Standards – KMLS) (for a full description of MTSK, including its singularity with respect to other models of teacher knowledge, see Carrillo et al., 2018)

It is the model in its entirety which is considered specialized. Consequently, our interest lies less in being able to identify items of knowledge within each subdomain, and more in the complex network of connections interlinking them in replication of the teacher's knowledge. To this end the model further divides each subdomain into categories. This enables it not only to operationalize analysis of teachers' specialized knowledge, but also to serve as a tool for critical reflection on the part of the teachers. Thus it is used in the PIC. The MTSK model enables members to reflect on their knowledge, typically taking as a starting point the need to plan problem-solving tasks for use in the classroom. This starting point leads to consideration of the distinct areas of knowledge which are brought into play, or which are needed to manage these tasks. The MTSK-guided reflection takes place at the planning stage and in the post-observation joint discussion of the implementation of the planned task.

In summary, in the PIC, teacher knowledge is key to the process of professional development. In the following section we present the specialized knowledge, as observed and identified by the researchers using the MTSK model, of one of the primary teacher-members of the group.

4. Enrique's MTSK

We now focus on one of the members of the PIC, a primary teacher named Enrique. His original training was in general education, with just a single course devoted to mathematics, amounting to 90 out of 1800 hours of the complete degree. At the time of this study, he had been teaching for some 20 years and did not usually teach mathematics. He brought a positive attitude to the PIC, happily taking on the responsibility for organizing sessions, managing proposals for areas of study and activities and planning lesson delivery. In this instance, he volunteered to carry out a lesson planned around how to define a polygon.

It is generally accepted that the mathematical knowledge of a primary teacher is less extensive than that of a secondary teacher. Evidently, there is no comparison between the amount of mathematics each studies over the course of their respective degrees, as secondary teachers do at least 240 credits (1800 hours) exclusively (or nearly exclusively) devoted to mathematics. It is also the case that Enrique, who would be unable to provide solutions to problems involving differential equations or advanced topology, is nevertheless well-equipped in the essential aspects of everyday mathematics beyond the content items studied in primary lessons. It is noticeable that Enrique highlights how you can do mathematics at the primary level in such a way that these fundamentals of everyday mathematics enter into circulation in the classroom discourse.

Enrique teaches the 5th year (age 10) at a state primary school in a small town on the coast in the province of Huelva (Spain). There are 25 pupils in his class, of varying levels and from diverse socio-economic backgrounds. In the lesson extract that we will analyse, he aims to guide his pupils towards reaching a definition of polygon. In point of fact, the pupils have already studied polygons in their 4th year, but Enrique is aware that they need to do more work on mathematical definitions and believes they should be capable of arriving at a meaningful definition for themselves, in particular to think of and test criteria on which to base their classifications.

The members of the PIC had discussed various ways of approaching the definition of a polygon with the 5th-year pupils, bearing in mind that they had already studied polygons in the 4th year. Some members suggested starting from the definition that they had met in 4th year and then moving on to an open-ended activity on classifying polygons, that is, without providing any criteria for the classification. The group agreed that the notion of classification was a process that went beyond specific mathematics classifications (in this case, the classification of polygons), and hence it would be appropriate that the pupils learn to classify. However, Enrique preferred not to assume that the pupils had assimilated the definition given in 4th year, as they had not spent much time on developing the definition. Instead their 4th-year teacher had only drawn the pupils' attention to the definition in the textbook. For this reason, Enrique suggested planning an activity in which the pupils construct a definition of a polygon. By doing so, Enrique thought, in consonance with the other members of the PIC, his pupils would go beyond merely understanding the definition of a polygon to jointly construct it, and that this would give them useful experience in how elements are defined in mathematics.

In the follow-up interview to this lesson, he says:

E (ENRIQUE): Even when they give you a correct definition, they don't have all the possible polygons in mind. What's more, they rarely get to grips with a proper mathematical definition; instead they have a set of properties which might be enough to define a polygon or might not be.

Various elements of MTSK are brought into play here. In the first part of his utterance, Enrique demonstrates his awareness of his pupils' knowledge gap between the definition of polygon and the set of polygons to which that definition corresponds (KFLM in relation to students' learning difficulties), as sometimes happens when a pupil fails to recognize a concave quadrilateral as a polygon despite the fact that it fulfils all the requirements of the definition. Further, Enrique gives sense of some awareness of necessary and sufficient conditions for a definition in mathematics (KPM in relation to the necessary and sufficient conditions of a definition). Finally, Enrique's insistence on the pupils developing a meaningful definition is a demonstration of his beliefs about mathematics teaching and learning.

In the lesson itself Enrique brings a bag to class filled with different flat shapes cut out of cardboard. He invites various pupils to each take a shape from the bag and stick it on the board. He instructs them to divide the shapes into two groups, one on the left and one on the right, but gives no further instructions and no indication as to the rationale they should use to form the groups. Nevertheless, those on the left correspond to polygons (including concave polygons) and those on the right to non-polygons (shapes with partial or total curved outlines, including a circle) (Figure 6.2). In this respect, Enrique displays KMT regarding the underlying design of this activity (with support from the PIC, naturally). He has clearly considered a wide variety of flat shapes for his pupils to sort into two groups with the awareness that in doing so the pupils would necessarily have to consider what features are common to each group, something fundamental for developing definitions (and again connected to his KPM).

Following this initial phase, Enrique then tells the pupils to look carefully at the group on the left and asks if they can remember a name to describe them. After a few moments of pondering, one of the pupils replies that they are called polygons, upon which Enrique tells the class that their task is now to define what a polygon is, and in order to do this, they need to focus on the features common to the set of shapes in each group. When a pupil suggests that one of the features common to polygons is that they have corners, Enrique observes that although this is true, there are also shapes in the other group which have corners. He adds:

E: When we define something, we try to find the common features, but in such a way that we also exclude the shapes that don't have all the features.

FIGURE 6.2 Starting to define a polygon (constructing the definition).

Here we can see evidence for Enrique's KPM regarding the features of mathematical definitions.

The lesson continues with Enrique writing on the board the features suggested by the pupils. In order to ensure that the figures on the right are excluded, he writes a negative feature: polygons do not have curves. He goes on to draw a shape (an open polygonal chain) which fulfils all the features the pupils have provided up to this moment, intending that the pupils should reject it from the set of polygons (as they subsequently do) on the grounds that polygons are closed shapes (Figure 6.3). He then adds this feature to the list on the board. In this instance, in addition to KPM, Enrique also demonstrates KoT, with respect to the topic of polygons (definition and properties).

In the next stage of the lesson, Enrique then draws a convex polygon on the board and asks whether the polygon is constituted by the line or what is inside the line (Figure 6.4). In the follow-up interview we asked what his reasons were for asking this question:

E: Well, they need to know the difference between a polygon and its outline. Up to now we have just talked about the outline; in the definition we've only mentioned features relating to the sides. I know that a lot of the pupils are not sure about this.

FIGURE 6.3 Prompting pupils to include the criterion of "closed" in the definition.

FIGURE 6.4 Differentiating border and interior.

In this instance, two aspects of MTSK come together: KoT (definition of a polygon and appropriate examples) forms a connection with KFLM (knowledge of typical areas of student difficulty).

We found evidence of Enrique's knowledge with respect to these four subdomains (KoT, KPM, KMT, KFLM) and, additionally, evidence of his beliefs about teaching and learning mathematics. Evidence of the other two subdomains (KSM and KMLS) was not found in this particular excerpt.

It is also worth noting the evidence Enrique displays of KPM. It is not common to observe activities in primary and secondary lessons dedicated to discussing with the students mathematical processes like making definitions. In most cases teachers base their work on definitions on memorization without exploring the underlying meaning and making it impossible to consider the characteristics of the process. This approach is independent of the educational level. Although a secondary teacher has wider training in maths, this by no means ensures that the teacher will provide their pupils with opportunities for reflection on how mathematics is done. With more limited mathematical knowledge, as mentioned earlier, Enrique was able to carry out a teaching activity which surpassed the specifics of defining a polygon to consider the process of giving definitions. His pupils got the opportunity to engage with genuine mathematical thinking by grappling with one of its most emblematic processes. He required not that they recite a definition by heart, but instead that they attempt to understand what conditions must be met for their jointly constructed formulation to be considered a definition.

Once again, the role of the collaboration of the group (the PIC) is instrumental. It is not a question of whether Enrique would have been able to plan the activity and put it into practice alone without the support of the PIC. It is rather a question of how the PIC made it possible, and how in addition it facilitated growth in members' knowledge and hence their professional development. This is achieved because, in the first place, the PIC is a non-threatening environment governed by good-humoured fellow feeling among its members. Never does anyone feel they are being judged. Instead, members share reflections with the aim of improving professional growth all round. Critical analysis, then, is always constructive. In the second place, the PIC's modus operandi is reflection. Individual and collective reflection are the hallmarks of what takes place during and between sessions. In the third place, the educational activities are oriented towards exploring and solving problems, as the group is convinced that this is the most effective way to learn mathematics. In the fourth place, the aim of the PIC is professional development. All members are willing to make all necessary changes to the dynamics and content of their work in order to further their professional development. On this point it is important to differentiate the teachers and the researchers. While the teachers are generally interested in their professional development, the researchers, in addition to sharing the teachers' interests, are keen to discover the keys or distinguishing features of such development. However, the teachers are also aware that research (based on a reflective approach) into their practice is a fruitful means to their professional development.

These defining features are the key to understanding the role the group takes in the professional development of its members. To these we can add that – at the current time – reflection on the knowledge teachers bring to bear on lesson activities is at the forefront of the PIC's methodology, and the reason we can confidently assert that the group has deepened the knowledge of Enrique and the rest of the members. To illustrate this, we can take a few extracts from the meetings in which Enrique's performance is analysed by the group (from a video recording, from which the photographs in the foregoing figures are taken).

The first example concerns the question which Enrique asked his pupils about whether the polygon is "what's inside" or "just the outline".

ROCÍO (pre-school teacher): What that boy said made me laugh, that sometimes the polygon was what is inside but other times it was the outline. And I liked how Enrique responded when he said that wasn't possible and that it always had to be either the inside or the outline.

INMA (primary teacher): Absolutely. Enrique has taught them that it doesn't make any sense to be able to vary a definition because then we wouldn't be able to understand each other.

This exchange between two of the teaching members of the PIC introduces into the debate elements of KPM, namely the importance of invariance in definitions in mathematical discourse – "invariance" here not implying that alternative definitions are not available.

Although the activity was originally planned in a PIC session, another teacher, on reflection about what happened in the lesson, suggested an adaptation to the material which was actually used:

ANA (primary teacher): I think the edges of the cardboard shapes could have been highlighted. Then you'd have been able to clearly distinguish the edges from the inside area.

ENRIQUE: I think that might have made things worse.

R (researcher): Why do you think doing that would make the association of the notion of a polygon with a flat area worse?

ENRIQUE: The cardboard shape represents a flat area, and that is the polygon. However, if we highlighted the edges, the pupils might think that the polygon is the edges, reinforcing the wrong idea that some children have because when we draw it on the board, what we draw are the edges.

Ana's utterance leads to reflect on the teaching resource. Enrique's reply shows KMT and KFLM, in that, in addition to understanding the essential properties of the materials (the cardboard shapes), he is able to connect this to the kind of error his pupils frequently make.

The second example focuses on reflection on the definition of a polygon which Enrique's class finally arrived at (Figure 6.5). The definition is the following: "All the shapes have straight sides, they have angles, they have vertices, they don't have any curved lines, are flat and all the sides have to be connected at their ends". For most of the members of the PIC, this definition is appropriate to primary level as they feel the pupils are ready to include a list of properties for delineating a concept, but are not ready to sift through the list so as to pare it down to the minimum number of properties necessary.

The teachers demonstrate their knowledge of what can be expected at a particular educational stage (KMLS), but at the same time certain difficulties can be noted regarding their understanding of how definitions are made in mathematics (KPM):

MARIO (secondary teacher): You don't need to say they have angles and vertices.
ANA: Why not? Polygons have angles and vertices.
MARIO: Yes, but if you specify that there are sides which all join up at the ends, you automatically get angles and vertices.

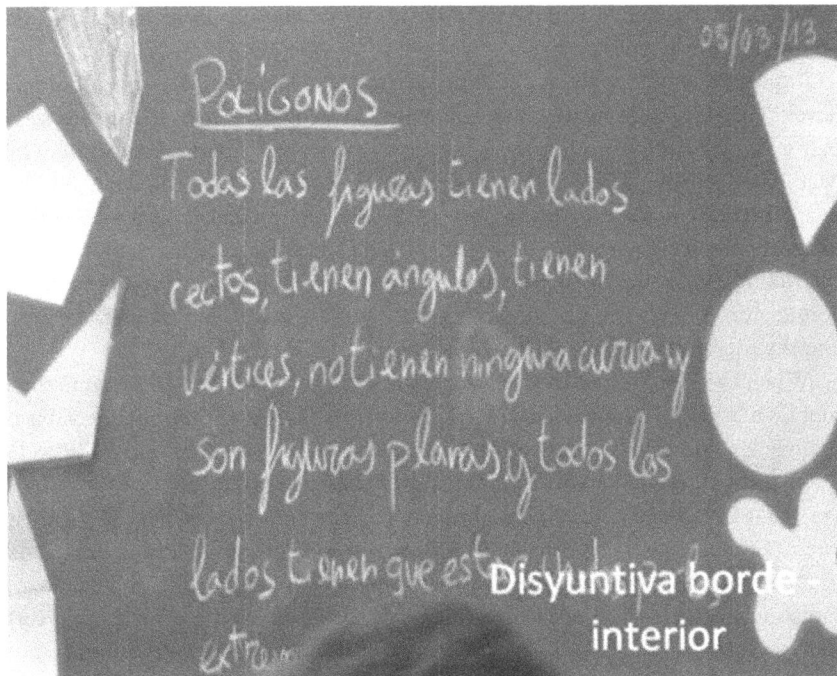

FIGURE 6.5 Definition of a polygon after the differentiation between border and interior.

JUAN (primary teacher): Well, OK, but it doesn't hurt to say so, does it?

MARIO: Definitions shouldn't really include properties which can be deduced from others.

This exchange encapsulates an area of knowledge, ostensibly shared by Enrique, concerning particular features of mathematical definitions. Judging from the lesson observation, for Enrique, all properties characterizing the item should be included in the definition so as to clearly exclude any other item which does not meet this set of properties. However, it would seem that, although Enrique is aware of the need for getting a sufficient set of conditions, he is unaware of the minimalist feature of mathematics definitions. The debate about these issues, involving all the members, results in valuable information about mathematics practices regarding definitions entering the discourse.

5. Applications of specialized knowledge

The PIC's deliberations about the keys to professional development, like the researchers' explorations, are not regarded by the group as definitive. More than theoretical speculations, they are seen as a means of creating the right conditions for professional development to flourish, and a point of departure for planning suitable activities and learning environments for facilitating meaningful learning on the part of the pupil.

Expressed another way, it could be said that the key elements of professional development – and in its way, the mathematics teacher's specialized knowledge – play a significant role in improving the quality of mathematics education. The Mathematical Quality of Instruction (MQI), developed by Hill et al. (2008), is an instrument for measuring various teacher competences in mathematics, including aspects such as how teachers respond to pupils (especially pupil error), the richness of the mathematics (especially in terms of reasoning and proof), the connection between the lesson activity and the core concepts in mathematics, and the use of mathematical language.

When the members of the PIC discuss Enrique's MTSK and at the same time mobilize a shared knowledge, the group affirms how it is that our specialized knowledge equips us for delivering quality education.

In this respect it is worth taking a moment to consider one of the teacher's out-of-class tasks, that of speaking to parents or other teachers about their work. Often teachers are unable to access a precise professional discourse, and in consequence their communication with parents is scattered with generalizations such as their child needs to study more, or he or she makes a lot of errors in their calculations. By the same token, conversations between teachers sometimes lack the sharp-edged precision that in-depth knowledge confers and are instead conducted through fuzzy approximations. By contrast, a teacher with a good grounding in knowledge of the features of learning mathematics will be able to talk to parents about the exact nature of their child's errors, of the kind of difficulties they may be having with certain types of mathematics problems. Likewise, assuming an equivalent knowledge of mathematics teaching, he or she can talk authoritatively with other teachers about the best strategies for helping the pupil overcome these difficulties. In short, as the teacher develops within the profession, so a gap widens between their discourse and that of any educated person on the topic. The PIC, as an environment of professional development, contributes to this growth in a similar way to that narrated by Wilson, Sztajn, Edgington, Webb and Myers (2017) about teachers' discussions on learning trajectories.

Returning to the classroom, Enrique can also be seen to adjust the way he interacts with his pupils according to his understanding of how they best learn and where they tend to find difficulties (KFLM), an expertise which enables teachers to deal with unforeseen situations that might arise. Clearly, KFLM is not the only subdomain which teachers draw on in such situations; KoT also plays a significant part in such responses. Although the PIC sessions have never focused directly on the teachers' knowledge of mathematics, instead broaching it indirectly through teaching scenarios, the teachers have considered this knowledge in terms of planning *substantial learning environments* (Wittmann, 2001), one of the features of which is the richness of the mathematics involved. It was in this spirit that some time ago the practising teachers in the PIC decided to tackle methods for division in the 3rd year of primary school. The lack of intuitive sense which their pupils tended to find in the traditional method was a cause of concern to them. The PIC provided them with an opportunity to discuss the best examples for introducing the topic, the materials and resources available for support, the different types of division (partitive versus quotitive) and ways of approaching the remainder, all of which not only supplied them with activities for easing the learning process ("like they'd never seen before", according to reports), but also generated an environment rich in professional development in which elements from various subdomains of MTSK were mobilized. Of particular note was their progress in terms of formulating examples with the considerations of variety and transparency (Bills et al., 2006) always in mind.

Alongside the feeling of belonging to a group, the knowledge gained by the teachers in the PIC provides them with security (they are now more willing to try out more open-ended activities than they were before joining the group) and currently empowers them to apply a *what if* (Brown & Walter, 1990) strategy. Applying this kind of strategy implies making connections (KSM) and being prepared to explore one's own mathematical knowledge. It means, too, being willing to formulate problems and being able to take a critical approach to the use of the textbook.

6. Final observations

By and large, relations between teachers and researchers can sometimes be difficult. Teachers tend to regard researchers as out of touch with the day-to-day realities of the classroom and are not convinced that the results of our research are applicable to their field. Conversely, some researchers suspect that teachers are interested only in instant solutions to immediate problems. Collaborative environments like the PIC underline that another kind of working relationships between teachers and researchers is possible (Feldman, 1993), in which the interests of both parties are respected. In such environments, teachers and researchers need to be willing to give generously but also need to feel rewarded for their participation. Teachers need to see tangible benefits to their teaching; researchers need to see the opportunity to contribute theoretical instruments to the realm of mathematics education. Both are committed to a process of professional development, and each has their personal objectives, respected by the others, as well as shared objectives which the group collectively pursues. The two parties each become researchers into their own practices and have an interest in the development of the others. For the researchers, working in the PIC has been a central pillar in constructing the MTSK model. For their part, the teaching members of the PIC continue their professional progression with an emphasis on the development of their specialized knowledge. The way they frame questions for their pupils has changed, as has the way they respond to student error, introduce concepts and procedures and plan the construction of definitions, along with their ability to formulate examples, their awareness of manipulatives and technological resources and their approach to the textbook. Their ability to guide constructive discussion in class and to deal with unforeseen events has grown, alongside their deployment of investigative activities. In summary, through a process of reflection emphasizing the construction of specialized knowledge with the chief motivation of becoming more competent (Hošpesová & Tichá, 2006) in managing problem-solving activities in class, the practising teachers in the PIC have gained elements which improve the quality of their mathematics instruction. Witnessing their pupils, under their guidance, construct mathematical arguments appropriate for their level – effectively introducing mathematical reasoning into their lessons – is a gratifying experience for both the teachers and the researchers. It is particularly rewarding for the researchers to see how learning opportunities are created for teachers, researchers and students alike, and that the teachers acquire competences which promote the sustainability of such development (Zehetmeier, 2010), when dynamic collaborative relations, teachers committed to their professional development and specialized knowledge intersect.

The teaching members of the PIC have, in short, developed a sense of themselves as mathematics teachers, significantly changing their professional identity (especially the primary teachers) (Wenger, 1998). We researchers have been motivated by the opportunity to explore pertinent areas of mathematics education in direct association with the needs of the teachers. Likewise, we have witnessed throughout the process of professional development the successful incorporation of elements such as the involvement of the teachers in decision-making (Clarke, 1994), evidence of consistency and gradual growth, including the need for support outside the school (Hawley & Valli, 1999), and a progression in the complexity and quality of the teachers' reflections on classroom episodes and their own knowledge (Carrillo & Climent, 2011).

Acknowledgement

The Spanish Government (EDU2013–44047-P, RTI2018-096547-B-I00 and EDU2016–81994-REDT) and the Research Group DESYM (funded by the Andalusian Government and the University of Huelva) supported this research.

References

Ball, D. L., Thames, M. H., & Phelps, G. (2008). Content knowledge for teaching: What makes it special? *Journal of Teacher Education, 59*(5), 389–407.

Bills, L., Dreyfus, T., Mason, J., Tsamir, P., Watson, A., & Zaslavsky, O. (2006). Exemplification in mathematics education. In J. Novotna (Ed.), *Proceedings of the 30th Conference of the International Group for the Psychology of Mathematics Education* (Vol. 1, pp. 126–154). Prague, Czech Republic: PME.

Brown, S. I., & Walter, I. (1990). *The art of problem posing* (2nd ed.). Hillsdale, NJ: Lawrence Erlbaum Associates.

Carrillo, J. (1999). Conceptions and problem solving: A starting point and a tool for professional development. In N. Ellerton (Ed.), *Mathematics teacher development: International perspectives* (pp. 27–46). West Perth, Australia: Meridian Press.

Carrillo, J., & Climent, N. (2011). The development of teachers' expertise through their analyses of good practice in the mathematics classroom. *ZDM Mathematics Education, 43*(6–7), 915–926.

Carrillo, J., Climent, N., Montes, M., Contreras, L. C., Flores-Medrano, E., Escudero-Ávila, D. . . . Muñoz-Catalán, M. C. (2018). The Mathematics Teacher's Specialised Knowledge (MTSK) model. *Research in Mathematics Education* (Online). doi:10.1080/14794802.2018.1479981

Carrillo, J., & Contreras, L. C. (1994). The relationship between the teachers' conceptions of mathematics and of mathematics teaching: A model using categories and descriptors for their analysis. In J. P. Ponte & J. F. Matos (Eds.), *Proceedings of the 18th Conference of the International Group for the Psychology of Mathematics Education* (Vol. 2, pp. 152–159). Lisbon, Portugal: PME.

Charmaz, K. (2014). *Constructing grounded theory* (2nd ed.). London, UK: Sage Publications.

Clarke, D. (1994). Ten key principles from research on the professional development of mathematics teachers. In D. B. Aichele & A. F. Coxford (Eds.), *Professional development for teachers of mathematics* (pp. 37–48). Reston, VA, USA: National Council of Teachers of Mathematics.

Climent, N. (2005). *El desarrollo profesional del maestro de Primaria respecto de la enseñanza de la matemática. Un estudio de caso* [The professional development of primary teachers with respect to the teaching of mathematics: A case study] (Doctoral dissertation). Proquest Michigan University, Michigan, US. Retrieved from www.proquest.co.uk

Climent, N., & Carrillo, J. (2002). Developing and researching professional knowledge with primary teachers. In J. Novotná (Ed.), *Proceedings of the Second Congress of the European Society for Research in Mathematics Education, CERME2* (pp. 269–280). Prague, Czech Republic: Charles University.

Contreras, L. C., Carrillo, J., & Guevara, F. (1996). The teacher as problem solver and his/her conception on problems solving in the mathematics classroom. In L. Puig & A. Gutiérrez (Eds.), *Proceedings of the 20th Conference of the International Group for the Psychology of Mathematics Education* (Vol. 1, p. 209). Valencia, Spain: PME.

Ernest, P. (1991). *The philosophy of mathematics education*. London, UK: The Falmer Press.

Feldman, A. (1993). Promoting equitable collaboration between university researchers and school teachers. *Qualitative Studies in Education*, *6*(4), 341–357.

Goos, M. (2014). Researcher: Teacher relationships and models for teaching development in mathematics education. *ZDM Mathematics Education*, *46*, 189–200.

Grbich, C. (2013). *Qualitative data analysis: An introduction*. Thousand Oaks, CA: Sage Publications.

Hawley, W. D., & Valli, L. (1999). The essentials of effective professional development: A new consensus. In L. Darling-Hammond & G. Sykes (Eds.), *Teaching as the learning profession: Handbook of policy and practice* (pp. 127–150). San Francisco, USA: Jossey-Bass.

Hill, H. C., Blunk, M. L., Charalambous, C. Y., Lewis, J. M., Phelps, G. C., . . . Ball, D. L. (2008). Mathematical knowledge for teaching and the mathematical quality of instruction: An exploratory study. *Cognition and Instruction*, *26*(4), 430–511.

Hospesová, A., Carrillo, J., & Santos, L. (2018). Mathematics teacher education and professional development. In T. Dreyfus, M. Artigue, D. Potari, S. Prediger, & K. Ruthven (Eds.), *Developing research in mathematics education: Twenty years of communication, cooperation and collaboration in Europe* (pp. 181–195). New Perspectives on Research in Mathematics Education Series, Vol. 1. Oxon, UK: Routledge.

Hošpesová, A., & Tichá, M. (2006). Developing mathematics teacher's competence. In M. Bosch (Eds.), *Proceedings of the Fourth Congress of the European Society for Research in Mathematics Education, CERME4* (pp. 1483–1493). Barcelona, Spain: FUNDEMI IQS, Universitat Ramon Llull.

Jaworski, B. (2005). Learning communities in mathematics: Creating an inquiry community between teachers and didacticians. *Research in Mathematics Education*, *7*(1), 101–119.

Keffer, A., Wood, D., Carr, S., Mattison, L., & Lanier, B. (1998). Ownership and the well-planned study. In B. S. Bisplinghoff & J. Allen (Eds.), *Engaging teachers: Creating teaching/researching relationships* (pp. 27–34). Portsmouth, NH, US: Heinemann.

Lubinski, C. A., & Vacc, N. N. (1994). The influence of teachers' beliefs and knowledge on learning environments. *Arithmetic Teacher*, *41*(8), 476–479.

Ma, L. (1999). *Knowing and teaching elementary mathematics: Teachers' understanding of fundamental mathematics in China and the United States*. Mahwah, NJ: Lawrence Erlbaum Associates.

Ponte, J. P., & Chapman, O. (2006). Mathematics teachers' knowledge and practices. In A. Gutierrez & P. Boero (Eds.), *Handbook of research on the psychology of mathematics education: Past, present and future* (pp. 461–494). Rotterdam, The Netherlands: Sense Publishers.

Rowland, T., Turner, F., Thwaites, A., & Huckstep, P. (2009). *Developing primary mathematics teaching: Reflecting on practice with the knowledge Quartet*. London, UK: Sage Publications.

Scheiner, T., Montes, M. A., Godino, J. D., Carrillo, J., & Pino-Fan, L. R. (2019). What makes mathematics teacher knowledge specialized? Offering alternative views. *International Journal of Science and Mathematics Education*, *17*(1), 153–172. doi:10.1007/s10763-017-9859-6

Shulman, L. S. (1986). Those who understand: Knowledge growth in teaching. *Educational Researcher*, *15*(2), 4–14.

Wenger, E. (1998). *Communities of practice: Learning, meaning and identity*. New York, USA: Cambridge University Press.

Wilson, P. H., Sztajn, P., Edgington, C., Webb, J., & Myers, M. (2017). Changes in teachers' discourse about students in a professional development on learning trajectories. *American Educational Research Journal, 54*(3), 568–604.

Wittmann, E. C. (2001). Drawing on the richness of elementary mathematics in designing substantial learning environments. In M. Van den Heuvel Panhuizen (Ed.), *Proceedings of the 25th Conference of the International Group for the Psychology of Mathematics Education 1 (PME 25)* (pp. 193–197). Utrecht, The Netherlands: Utrecht University.

Zehetmeier, S. (2010). Sustainability of professional development. In V. Durand-Guerrier, S. Soury-Lavergne, & F. Arzarello (Eds.), *Proceedings of the Sixth Congress of the European Society for Research in Mathematics Education, CERME6*. Lyon, France: INRP.

7

THE SPECIALIZED KNOWLEDGE AND BELIEFS OF TWO UNIVERSITY LECTURERS IN LINEAR ALGEBRA

Diana Vasco Mora & Nuria Climent Rodríguez

1. Introduction

There can be little doubt that teacher training plays a critical role in the efforts to construct an effective system of mathematics education. If we want the best outcomes for our students' learning, we must necessarily take a critical interest in our teachers, in what they know and in what they are capable of doing (Even & Ball, 2009). While recent research into mathematics teacher education has focused on mathematics teachers' knowledge, research into the knowledge of university lecturers has been underrepresented (Lin & Rowland, 2016). This area is in fact one of the lines of research which Biza, Giraldo, Hochmut, Kharkbaz and Rasmussen (2016) identify as a potential way forward for future research in their review of research into mathematics education at university level.

We scrutinize the knowledge deployed by two university lecturers in linear algebra. This focus is of particular interest if we consider the evidence available from research of the difficulties university students face in this particular area (Dorier, 2002). However, as Fukawa-Connelly, Johnson and Keller (2016) point out, there is a lack of research into lecturers' beliefs, knowledge and goals regarding the teaching of algebra at this level.

At the same time, there is evidence of a connection between teachers' knowledge and their beliefs, and that the interactions between these two constructs can help us to better understand the two (Charalambous, 2015).

In this study we explore the knowledge displayed by two university lecturers teaching the topic of *matrices, determinants and systems of linear equations*. In particular, we examine the potential connections between this knowledge and their beliefs about teaching and learning mathematics, with a view to understanding how such connections are used as resources and personal orientations in shaping their practice (Schoenfeld, 2010).

2. Theoretical framework

This study concerns the interconnectivity of mathematics teachers' knowledge and their beliefs about teaching and learning. For this purpose, we have opted to divide the theoretical framework into two sections, the first dealing with teachers' knowledge at secondary and university levels, and a description of the analytical model, the *Mathematics Teacher's Specialized Knowledge* (henceforth MTSK) (Carrillo, Montes, Contreras & Climent, 2017), and the second with mathematics teachers' beliefs about teaching and learning.

2.1 Mathematics teachers' knowledge

The categories proposed by Shulman (1986, 1987), have inspired the development of other models for characterizing teachers' knowledge for mathematics education at various levels, and even for specific courses. Worthy of note among these alternatives are *Mathematical Knowledge for Teaching* (MKT) (Ball et al., 2008) and *Mathematical Proficiency for Teaching* (Schoenfeld & Kilpatrick, 2008).

Specific to teachers' knowledge of algebra is the model developed by McCrory, Floden, Ferrini-Mundy, Reckase and Senk (2012), called *Knowledge of Algebra for Teaching* (KAT), with secondary and post-secondary teachers in mind. The model comprises three knowledge domains – *Knowledge of School Algebra, Knowledge of Advanced Mathematics* and *Mathematics for Teaching Knowledge* – and three categories of teaching practices – *decompressing* (deconstructing procedures and algorithms, and attaching fundamental meaning to symbols), *trimming* (knowledge of simplified versions of advanced concepts to be dealt with in subsequent courses) and *bridging* (which refers to efforts at making mathematical connections across topics, courses, concepts and objectives, including links between the ideas of school algebra and those of abstract algebra). The authors emphasize teaching practices and how these practices can support the emergence of categories of knowledge in the teachers' algebra classrooms.

In their study into a trainee secondary-level algebra teacher, Mamba, Mosvold and Bjuland (2017) note a significant influence of content knowledge on the prospective teacher's pedagogical content knowledge (in terms of his/her repertory of teaching strategies and his/her understanding of student errors).

In the context of university lecturers' knowledge of linear algebra, the categories of content knowledge and pedagogical content knowledge have been investigated. Regarding the latter, the ability to detect misconceptions on the part of the students has been identified as crucial, and the ability to give explanations which take into account the students' current stage of mathematical development, a skill requiring the ability to listen to students and make sense of their thinking (Johnson & Larsen, 2012).

In the case of our study, the next section describes the model we used to analyse the knowledge of two university lecturers in linear algebra.

2.1.1 The mathematics teacher's specialized knowledge

The instrument we use for examining teacher knowledge is the MTSK model (Carrillo, Climent, Contreras & Muñoz-Catalán, 2013; Carrillo et al., 2017), which focuses on knowledge unique to the mathematics teacher, and which understands specialization as a feature common to all domains making up the model (Figure 7.1).

Like all models of teacher knowledge, MTSK owes a debt to Shulman's (1986, 1987) seminal work but finds its direct origins in studies carried out using the MKT model (Ball et al., 2008) and the difficulties encountered with the demarcation of different subdomains therein (particularly Common Content Knowledge and Specialized Content Knowledge), when it was applied to teachers (Flores-Medrano, Escudero-Ávila & Carrillo, 2013), although the model recognizes that teachers' content knowledge substantively differs from that of other professionals who make use of mathematics. There are two knowledge domains in MTSK, Mathematical Knowledge and Pedagogical Content Knowledge, while beliefs are placed at the centre of the model as elements which permeate all areas of

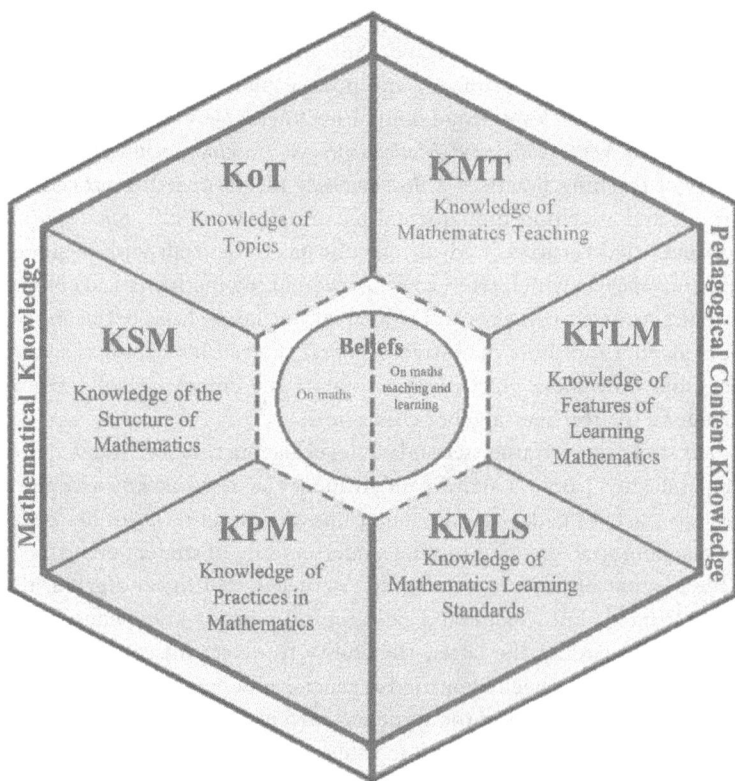

FIGURE 7.1 Domains and subdomains in the MTSK model.

Source: Carrillo et al. (2013).

knowledge. We give a brief description of the domains, subdomains and categories comprising the model; a full account can be found in Carrillo et al. (2018).

The domain Mathematical Knowledge is comprised of the subdomains Knowledge of Topics, Knowledge of the Structure of Mathematics and Knowledge of Practices in Mathematics, which are described next.

Knowledge of Topics (KoT) is the deep knowledge of mathematics content, including connections within the proximity of a single concept (*intraconceptual connections*) (Fernández & Figueiras, 2014). It is divided into four categories: *phenomenology and applications*, the knowledge of phenomena associated with the meanings of a mathematical topic, and its applications; *definitions, properties and foundations*, the knowledge for characterizing a concept and the properties of a mathematical object; *registers of representation*, or knowledge of the forms in which a topic can be represented, including notation and mathematical language; and *procedures*, itself comprising four subcategories: *how to do something*, *when to do something*, *why something is done this way* and *characteristics of the result*.

Knowledge of the Structure of Mathematics (KSM) refers to the interlinking systems which bind the subject. Four types of linkage are recognized: *connections based on simplification*, which trace content strands back to simpler treatments; *connections based on increased complexity*, which follow the strand forwards to more complex treatments; *auxiliary connections*, whereby one content element is used as a tool in the development of another content; and *transverse connections*, which are formed by the underlying ideas common to distinct mathematical content.

Knowledge of Practices in Mathematics (KPM) brings together knowledge of the ways of proceeding in mathematics, that is, the syntax of mathematics and the logic of the procedures in the discipline.

For its part, the domain *Pedagogical Content Knowledge* is divided into the subdomains Knowledge of Mathematics Teaching, Knowledge of Features of Learning Mathematics and Knowledge of Mathematics Learning Standards.

Knowledge of Mathematics Teaching (KMT), comprising knowledge of all aspects to do with teaching the subject, includes the categories *physical and digital teaching resources*; *theories of teaching* specific to mathematics education; and *strategies, techniques, tasks and examples* related to the topic in hand.

Knowledge of Features of Learning Mathematics (KFLM) includes knowledge of pupils' *strengths and weaknesses* in respect of different topics; *ways pupils interact with mathematical content*, that is knowledge of pupils' strategies and the language commonly used for approaching content items; *theories of learning*; and pupils' *interests and expectations* regarding mathematical items.

Finally, *Knowledge of Mathematics Learning Standards* (KMLS) concerns knowledge of the official syllabus (and its sequence) and aspects of knowledge derived from journals, research groups and professional associations.

Although such neatly parcelled-out models might give the impression of an atomistic approach to teacher knowledge, we are aware that an individual's knowledge is a holistic system. MTSK is designed such that, whilst it allows a fine-grained analysis of specialized knowledge, in this case that of two university lecturers, at the same time it recognizes the inter-relatedness of elements of knowledge, and the potential relationship these have with the teacher's beliefs about teaching and learning mathematics.

2.2 Teacher beliefs about teaching and learning mathematics

Researchers into mathematics education have shown an interest in teachers' beliefs for more than three decades (see, for example, Thompson's 1992 cornerstone review). The question of what can be understood by teachers' beliefs and in what way they are distinct from the construct of "teachers' knowledge" has been a source of controversy from the start, an issue which persists today (e.g. Törner, 2002; Fives & Buehl, 2012).

We understand beliefs as an individual's set of incontrovertible personal truths, incorporating, without distinction, beliefs, conceptions, mental images, concepts, meanings and preferences (Thompson, 1992). Despite the inherent epistemic problems associated with studying teachers' beliefs, we would say that taking into consideration a teacher's thinking about, say, what for him or her is important for teaching a specific item in a specific context can contribute to accounting for their decisions and actions.

In this study the teacher's beliefs are extrapolated largely from their practice. Rather than worry about possible inconsistencies between the teacher's declared beliefs and their actual practice, we assume coherence between beliefs and knowledge within what Leatham (2006) calls a sensible system. Hence, we consider the teacher's beliefs, like their knowledge, within the context and time frame of the practice from which they were inferred.

The influence of the teacher's beliefs on what takes place in class, and ultimately what the students learn, has been recognized for more than two decades (Wilson & Cooney, 2002). The teacher's knowledge and orientations (including his or her beliefs) influence their perceptions of classroom situations, and these perceptions are a predictor of teacher planning (Dunekacke, Jenßen, Eilerts & Blömeke, 2016).

We can say that at the university level, research has also confirmed that the beliefs of lecturers have a significant influence on classroom practice (Biza et al., 2016). However, there are few studies into university lecturers' conceptions. In fact, when Forgasz and Leder (2008) review the research literature into teacher beliefs about teaching and learning mathematics between 1997 and 2006, they make reference only to teachers at primary, middle years and secondary school. Despite a decade having passed since this review, the almost complete absence of studies remains (Fukawa-Connelly et al., 2016).

Various studies have found links between teachers' knowledge and beliefs. This relation has been described in various ways: knowledge (and lack thereof) feeding a resistance to a shift in beliefs (Britt, Irwin & Ritchie, 2001, in a study of middle school teachers); beliefs as mediators between knowledge and practice (Pajares, 1992) and potential interrelations between knowledge and beliefs (Clark et al., 2014). The latter study found that (with respect to upper-elementary teachers, but not middle-grade teachers) higher levels of mathematical knowledge correlated with less traditional beliefs about the role of students in mathematics lessons, whilst a negative correlation was obtained between teacher knowledge and the tendency to consider the teacher as the focus of the educational process.

The conceptions which have received most attention in research into practice are teachers' conceptions of mathematics, and the teaching and learning of mathematics, the latter constituting the focus of this study. Research with teachers at different stages of the educational system has highlighted that these beliefs are affected by a diverse range of factors and are dependent on context and students (Forgasz & Leder, 2008).

In many studies, conceptions about teaching and learning mathematics reflect models of teaching mathematics as they are defined in curricular reforms. It is thus often the case that two teaching methodologies are diametrically opposed: a transmission-oriented view (based on transmission theories of learning) and a constructivist view (emphasizing conceptual understanding and problem-solving) (e.g. Clark et al., 2014).

Carrillo and Contreras (1994), drawing on the typology of teaching styles identified by Kuhs and Ball (1986), distinguish four tendencies – *traditional, technological, spontaneous* and *investigative* – in teachers' conceptions of teaching and learning mathematics. Here, the traditional and investigative tendencies correspond to the aforementioned *transmission-oriented* and *constructivist* views respectively. The investigative tendency corresponds to the learner-focused view of learning of Khus and Ball (ibid.). The conceptual-oriented and performance-oriented views of Khus and Ball have a degree of overlap with the technological tendency (in terms of their utilitarian approach and the emphasis placed on understanding the logic of the discipline). The system used by Carrillo and Contreras (1994) for analysing teachers' conceptions consisted of six categories – methodology, subject significance, learning conception, student's role, teacher's role and assessment – and corresponding indicators for locating teachers' conceptions of these categories within one of the four tendencies. The descriptors drew on various previous studies (e.g. Kuhs & Ball, 1986; Ernest, 1991; Fennema & Franke, 1992). It is this system of categories which we selected in the study presented here for capturing the university lecturers' conceptions.

Of the few studies we were able to find dealing with the conceptions of university lecturers, the work of Fukawa-Connelly et al. (2016) deserves mention for its research into what pedagogical practices lecturers report using in their classrooms, and why. The authors find a strong tendency (82%) among those surveyed to rate the *lecture* as the most effective teaching mode, a result consistent with the 85% who stated that the mode most frequently used by them was *lecturing*. One of the main justifications for this preponderance of lecturing was the need to cover the course contents, which (at least at face value) would appear to be more a question of personal preference than any kind of externally imposed requirement. At the same time, with regard to learning, the majority (56%)

agreed that learning was most effective when they spent time in the class doing mathematics (besides listening to the lecture and taking notes). The authors note that these beliefs would appear to be mutually contradictory, particularly as 63% state that the students never solve mathematical problems in class, and that beyond taking notes, what they do is perform calculations and work on examples or applications. The lecturers cite their experiences as teachers (84%) and as students (64%) as the most influential factors in shaping their practice. On the other hand, in their study into the beliefs of two university lecturers about the role and use of mathematics in an engineering programme, Hernandes-Gomes and González-Martín (2015) note how the lecturers' beliefs cause them to approach topics such as mathematical rigor and approximation in a different way.

3. Methodology

The study took a qualitative approach (Neuman, 2014) with an instrumental case study design (Stake, 2003), focusing on two specific cases with the objective of characterizing the knowledge and beliefs of lecturers in linear algebra. It sought to answer the research question: What connections can be found between the specialized knowledge evident in the practice of two university lecturers in linear algebra and their beliefs about teaching and learning mathematics when they cover the topic of matrices, determinants and systems of linear equations? Our aim in this study was to deepen our understanding of mathematics lecturers in the technical undergraduate degrees, in particular with regard to how the knowledge and beliefs that are brought into play in the classroom help to shape their practice. Characteristics such as depth–detail, completeness and within–case variance are among the strengths of case study (Flyvbjerg, 2011), whilst the kind of generalization sought by an instrumental case study is here limited by the context and particularities of the cases studied.

3.1 The cases

The study was carried out at an Ecuadorian university with the collaboration of two lecturers, who for the purposes of this chapter we shall refer to as Jordy and Carlos, giving classes in the first year of an engineering degree. The linear algebra module was 16 weeks in length, with four hours of classes per week and an average of 20 students in each class. The course programme had been established previously and the topic in question, matrices, determinants and systems of linear equations, represented the point of departure for the linear algebra course and was the foundation for other content areas such as vectors in R^2, R^3 and R^n, vector spaces and linear transformations, to be covered later in the course. The lecturers were chosen for their willingness and interest in collaborating in the study, and for the convenience resulting from the fact that they were both members of the teaching corps to which the first author also belonged.

Jordy was a graduate in educational sciences, specializing in mathematics. He began his professional life as a mathematics teacher at a secondary school in 1993, a post he still held at the time of the study. In addition, he had been lecturing at the university for 9 years, the previous 6 of which he delivered a course in linear algebra for one of the engineering degrees. In the first 3 years, he taught a mathematics module on the introductory courses for new entrants to engineering degrees.

For his part, Carlos had studied geology and began his professional life as a cartography teacher. He became a university lecturer in 1998, teaching courses related to mathematics (algebra, trigonometry and analytical geometry in the Faculty of Animal Sciences). At the time of the study, he had been teaching linear algebra in various engineering degrees.

Both had followed courses in mathematics teaching (Jordy, short courses in teaching and learning mathematics at the secondary level; Carlos, a diploma in university teaching with emphasis on mathematics). Jordy held a master's degree in environmental science, Carlos in teaching. Both were at the time pursuing a doctoral degree in teaching.

The programmes of study (syllabi) at the university in which Jordy and Carlos worked were developed by the team of coordinators for each degree and sent to the lecturers teaching the modules (in this case, linear algebra) each semester. Then, as now, teachers were free to made certain modifications to the programme (tasks, practicals, additional content, bibliographic recommendations), as long as the minimum requirements in terms of content were maintained.

3.2 Data collection

The data were collected by means of video recordings (non-participatory observations) and semi-structured interviews over two consecutive teaching periods: November 2011–January 2012 (year 1) and November 2012–January 2013 (year 2). The aim was to gather sufficient information to enable an in-depth study of the teachers' knowledge and beliefs. The study formed a part of a broader research project, some of the results of which have already been published (e.g. Vasco & Climent, 2016; Vasco, Climent, Escudero-Ávila, Montes & Ribeiro, 2016).

In total, 13 of Jordy's class sessions and 7 of Carlos's were recorded. Three semi-structured interviews were also carried out with each lecturer, in order to obtain supplementary information on the lecturers' knowledge and beliefs, and to validate certain interpretations the researchers had made on the basis of the video analysis.

3.3 Data analysis

In order to effect the analysis of the data, we first transcribed the recordings of the videos and the interviews and then carried out a content analysis, an analytical technique using systematic procedures for describing content (consisting of pre-analysis, exploration of the material, treatment of results, inference and interpretation) (Bardin, 1996). In the first instance, we carried out a pre-analysis of the transcriptions of the video recordings, bearing in mind throughout the MTSK subdomains and categories, which gave us a broad perspective of the knowledge displayed by the actions of the teachers. The interviews with the lecturers were then analysed, and this was followed by a joint analysis of the recordings in order to confirm or reject the inferences that had been drawn about the lecturers' knowledge based on what they did and said in the course of their teaching. Through this means we were able to reach a definitive analysis and interpretation of their knowledge. The process involved a reduction in the raw data, as representative units of information were selected, categorized and coded according to the analytical framework.

The lecturers' beliefs about teaching and learning mathematics were analysed according to the four tendencies (*traditional, technological, spontaneous* and *investigative*) outlined in Carrillo and Contreras (1994), as described in the aforementioned theoretical framework, across the categories *methodology, subject significance, learning conception, student's role, teacher's role* and *assessment*. The descriptors for each category enabled the lecturers' beliefs to be located within the tendencies on the basis of their class performance and statements. For example, in the category *methodology*, the following descriptors are used to describe the teacher's practice: "class activity is characterized by the repetition of exercises" (*traditional tendency*); "the exercises aim to reproduce the logical processes explained in class, and to analyse student errors" (*technological tendency*). The intention here is not to expand upon the teaching tendencies. Rather, the framework acts as an aid to locating the beliefs of Jordy and Carlos in terms of their performance in class so as to be able to identify connections with their specialized knowledge.

4. The lecturers' knowledge and beliefs

Here we present two sections detailing the knowledge and beliefs of Jordy and Carlos based on the analysis of class excerpts (episodes) and interview extracts, which enable us to establish possible connections between their knowledge and beliefs and at the same time achieve an understanding of these lecturers' practices. In the episodes and interview extracts, J (Jordy) and C (Carlos) refer to each of the lecturers, and S to any contributing student. The episodes have been selected in function of the relevance of their content in illustrating the potential connections between knowledge and beliefs.

4.1 Jordy

There is abundant evidence of Jordy's knowledge of procedures during the course of his teaching the topic of matrices, determinants and systems of linear equations. With respect to determinants, we observed his knowledge of different calculation methods, including Sarrus's rule, the minor and cofactors method, properties and elementary row operations (KoT − *procedures*). The following is an example of an episode in which he calculates the determinant of a matrix by means of properties:

J: Let's start with properties which result in a determinant of zero (a student refers to when there are two equal rows). . . . It can be two rows or two col-

umns, like this $\begin{vmatrix} 3 & 4 & 3 \\ -1 & 5 & -1 \\ 2 & 7 & 2 \end{vmatrix}$ [the lecturer names other properties resulting in

a determinant of zero] . . .

In a matrix we can swap rows, and then the determinant changes its sign.

If we take this matrix $A = \begin{bmatrix} 2 & -1 \\ 3 & 4 \end{bmatrix}$ and we calculate its determinant, what's

the answer?

S: 11

J: Then we swap the rows and see what happens $|A| = \begin{vmatrix} 3 & 4 \\ 2 & -1 \end{vmatrix}$ which should

give what?

S: −11

J: Here, you need to be careful. When we work with matrices, we say that matrices are equivalent if we swap one row with another. But if we work with determinants and swap one row with another, the value of the determinant changes its sign. . . . Applying basic operations between rows, if you add to one row the product of another row multiplied by the number you want, the determinant is still the same. For example, let's change f2 for f2 +

2f1, we get $\begin{vmatrix} 2 & -1 \\ 7 & 2 \end{vmatrix}$. . . We need to learn this to make a triangular matrix

and calculate the determinant. With the same matrix A, do you remember how we do 5A?

S: It was like this $5A = \begin{bmatrix} 10 & -5 \\ 15 & 20 \end{bmatrix}$

J: How would you do it if you had $5|A|$?

S: You multiply just one row or by one column.

J: We have $5|A| = \begin{vmatrix} 10 & -1 \\ 15 & 4 \end{vmatrix} = 55$. When you multiply a determinant by a

scalar, you don't do it the same way as when you are dealing with a matrix where you multiply each element of the matrix one by one; instead you multiply just one row or one column. Matrices and determinants have different properties because they are two different things. . . . Be careful with the notation, a square bracket is not the same as bar. [The lecturer sets examples of determinants of order 3, 4 and 5 to calculate using their properties.]

Jordy's knowledge of different properties for calculating the determinant of a matrix can be noted in this extract (KoT – *definitions, properties and foundations*), alongside his awareness of potential areas of difficulty for the students (KFLM – *strengths and weaknesses*). He shows that he is aware that students tend to wrongly assume that certain properties germane to matrices can be generalized to determinants (viz., if a determinant swaps two rows, its value changes polarity, unlike what happens with matrices, which remain equivalent; also that to multiply a scalar by a determinant, the scalar is multiplied only by a row or column of the determinant), and students may not be aware of the difference in notation for distinguishing between matrices and determinants. This latter point also illustrates his knowledge of the algebraic-matricial register of representation (KoT – *registers of representation*). We can say that in this case, his knowledge of properties is subsumed by his knowledge of procedures (the use of the properties of determinants for making calculations) (KoT – *procedures, how to do something* and *why something is done in this way*).

In the extract, it can be seen how Jordy elicits the interaction of the students as he delivers his exposition, putting the emphasis on ensuring that they grasp the underlying reasoning and warning them of potential errors so that they don't fall victim to them. From this approach we infer that Jordy's belief about how to teach this content item is through exercises which replicate the logical processes involved and focus on (the avoidance of) student error. Indeed, the lesson observations registered his particular attention to common errors, such as when he immediately corrects a student who expresses the result of calculating a determinant between bars, the notation for absolute values (and hence inappropriate in this context), or when he corrects another student who (wrongly) puts the elements for calculating a determinant between brackets. These instances illustrate the importance Jordy gives to warning students about potential errors and rapidly correcting them when they occur, which connects his knowledge of this feature (KFLM) with his knowledge of the content area (KoT) and his belief about the role of errors in teaching.

The extract also illustrates how Jordy incorporates examples into his exposition of the contents, with the aim that they should then be able to replicate the reasoning involved. We can infer from this that he views teaching as a process whereby the lecturer expounds on content, not as an endpoint, but as something to be constructed with the support of examples. The knowledge of examples demonstrated by Jordy (KMT – *strategies, techniques, tasks and examples*), allied to his beliefs about how to use them, means that he not only states a property but also drives home the characteristics of each property and adjusts the examples to illustrate the distinct properties. Something of this tendency can be seen in the extract when he gives an example to show how a determinant can be zero if two columns, not just two rows, are equal. In addition, to illustrate the properties, he gives determinants of different dimensions and characteristics. This strategy is consistent with the *dimensions of possible variation* (Watson & Mason, 2005) in that it foregrounds such aspects as can vary or be modified in an example without ceasing to be an example of the topic in question, and also with *generalization* (Zaslavsky, 2010), in that it underlines those aspects of procedure which need to be illustrated, whilst at the same time indicating those which are arbitrary and modifiable.

After observing samples of Jordy's teaching, we can infer that his conception of learning tends towards a memory-based deductive model, the end of which is to impart information of a practical nature, and the means being the replication of exercises designed to fix in the minds of the students the procedures taught and their application to mathematics. This conception we connect with his knowledge of the applications of determinants (KoT – *phenomenology and applications*), in consonance with his views stated in one of the interviews, in which he explained how he drew his students' attention to the applications of the topic for solving systems of linear equations and inverse matrices, as corroborated by his observed practice.

Figure 7.2 illustrates a summary of aspects of Jordy's knowledge and beliefs which appear to be interconnected in respect of two seemingly key elements of his practice – the use of examples and the treatment of errors.

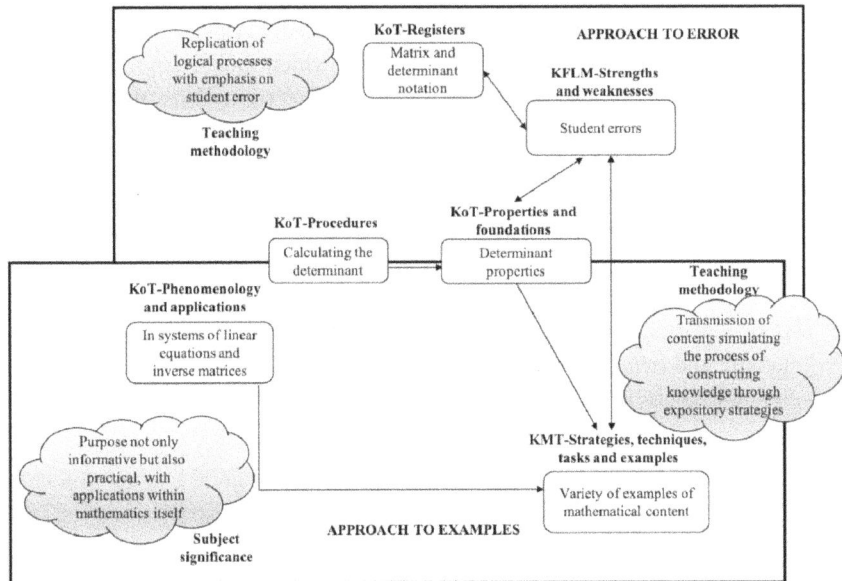

FIGURE 7.2   Associations between Jordy's knowledge and his beliefs.

4.2 Carlos

The following extract is a good illustration of Carlos's knowledge of the procedures for working with matrices, determinants and systems of linear equations:

C: We're going to find out the unknowns in a linear equation system, and to do so we first need to learn a method for calculating this determinant

$$\begin{vmatrix} -1 & 3 & 1/2 \\ 4 & 0 & 2 \\ -2 & -1 & 5 \end{vmatrix}$$ . . . [The lecturer explains the minor and cofactors method

for finding the determinant.] And afterwards we situate the final element

of the first row as cofactor . . . the signs are alternated $-1 \begin{vmatrix} 0 & 2 \\ -1 & 5 \end{vmatrix} -3 \begin{vmatrix} 4 & 2 \\ -2 & 5 \end{vmatrix}$

$+1/2 \begin{vmatrix} 4 & 0 \\ -2 & -1 \end{vmatrix}$ . . . To calculate the determinants, the diagonals are multi-

plied, taking into account that the result of the multiplication of the secondary diagonal is subtracted, and therefore, the sign changes. What is the value of the determinant?

S: −76

C: . . . We're going to look at some exercises for applying matrices. We'll solve

the linear equations system using the inverse matrix $A^{-1} = \dfrac{1}{|A|} \text{Adj}(A)$. We

have the following problem: In the first working week of a small company, 4 employees made 70 bunk beds, which needed 5 trips to be transported. In the second week, 6 employees made 80, which took 5 trips. And in the third week, 8 employees made 90, which took 6 trips. What should be the retail price of each bunk bed, the weekly wage of each employee and the cost of each transportation trip so that the earnings are $2700 for the first week, $2600 for the second and $2480 for the third? . . . In setting out the problem we need to identify the unknowns, which we'll convert into a matrix. What order?

S: Three.

C: Three equations, which assign positive values to anything that produces income and negative to anything that produces an expense. We'll separate the matrices from the coefficients, the unknowns and the earnings (labelled

C, S and G, respectively): $\begin{bmatrix} 70 & -4 & -5 \\ 80 & -6 & -5 \\ 90 & -8 & -6 \end{bmatrix} \times \begin{bmatrix} p \\ s \\ v \end{bmatrix} = \begin{bmatrix} 2700 \\ 2600 \\ 2480 \end{bmatrix}$. So, if you look

closely, here it says C × S = G, I isolate S and I'm left with S = C⁻¹ × G. What is C⁻¹? . . . [The students find the determinant and the lecturer explains how to obtain the adjugate and the inverse of a matrix.] You've found the determinant. . . . Whether it's positive or negative doesn't matter, what matters is that it has a value which is not zero so that the matrix has an inverse. . . . To find S, what do we do?

S: Multiply the inverse of C by G.

C: Exactly, so do that. . . . Then you get $\begin{bmatrix} 60 \\ 350 \\ 20 \end{bmatrix}$ where 60 should be the retail price

of the product, 350 the wage and 20 the cost of each trip for transportation.

Carlos's knowledge of the procedures for calculating the determinant via the minor and cofactors method can be observed here, along with setting up and solving linear equation systems with the inverse of the matrix (KoT – *procedures, how to do something*). To present the content, the lecturer first gives an exposition of the procedure to follow, giving an example to illustrate this, and then sets exercises to apply the technique. As such, it would seem that he regards student activity in classes as involving the repetition of exercises which closely reproduce the parameters of the examples (a view he corroborates in one of the interviews when he states that he selected the practice exercises for their similarity to those tackled in class), and presentation of content as an exposition by the teacher, giving illustrative examples to support the explanation, followed by the carrying out of exercises by the students (methodological beliefs).

In the extract, we can see that when Carlos makes use of an example in his exposition, his aim, in addition to clarifying the topic, is to illustrate how it can be applied to real situations (connected to the kind of professional work the students might be expected to meet in the future). In addition, Carlos calculates the determinant which will subsequently be used for solving the linear equation system, thus displaying in this case his knowledge of the applications of mathematical content within mathematics itself. In other words, his knowledge of how content can be applied (KoT – *phenomenology and applications*), of examples for teaching (KMT – *strategies, techniques and examples*) and his beliefs about mathematics education all intersect in his use of the examples.

Likewise, he explains the systems of linear equations using a problem which simulates a real situation. This bears out his knowledge of applications to everyday situations and relates to his beliefs about the purposes of the subject (viz. not solely the transmission of information, but of practical application to mathematics itself and to real life). Further evidence of these beliefs can be found in the following interview extract:

C: [I want] the students to realize that [linear algebra systems] have applications to real life. A lot of them think we study them for their own sake, but when you show them that they do have applications to real life, they show more interest in them.

Carlos associates these applications with keeping the students interested in the classes.

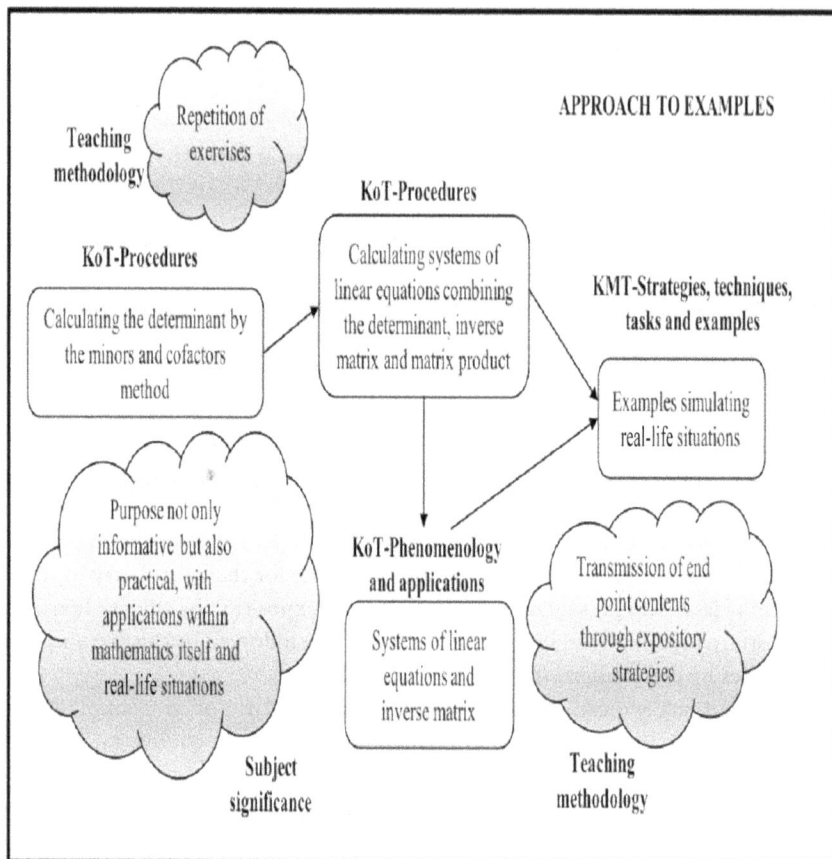

FIGURE 7.3   Associations between Carlos's knowledge and his beliefs.

Figure 7.3 provides a summary of the possible relations between Carlos's knowledge and his beliefs.

5. Discussion and conclusions

From close observation of the two lecturers' teaching matrices and determinants, we identified key facets of their practice (the role of examples in both cases, plus the role of errors in Jordy's case), which revealed a complex but consistent configuration of elements from their mathematical knowledge, pedagogical content knowledge and beliefs about the teaching and learning of mathematics. More specifically, in the case of Jordy, we can account for his emphasis on sign posting potentially problematic content in terms of the convergence of various aspects of his knowledge: his awareness of thorny areas (KFLM); his detailed knowledge of properties and registers of representation (both aspects of KoT); and his belief that teaching is based on setting exercises that replicate the required patterns of logic, with an emphasis on dealing with errors. In the case of Carlos, his knowledge of potential areas of difficulty is not evidenced in the observation of his practice, whilst the content knowledge he displays is concerned with procedure and the application of content to real situations (KoT – *procedures, how to do something* and *phenomenology and applications*).

Although both teachers seem to have the same conception of teaching, that is as a fundamentally expository activity in which they expound on the content with the support of examples, their use of examples differs. These differences, it seems to us, can be traced to the specialized knowledge and beliefs of each. In Jordy's case, he employs a variety of examples (KMT) which serve not only to illustrate the content, but also to alert students to potential errors. For this he draws on a range of knowledge – of *properties and foundations* (KoT), of areas of difficulty likely to cause errors on the part of the students (KFLM) and of how the content feeds into related mathematical procedures (KoT – *phenomenology and applications*) – and also on his belief that the subject matter serves a chiefly informative purpose and is chiefly practical in nature, focused on its applications to mathematics itself. Carlos's examples centre on the application of the content to situations which replicate as closely as possible authentic professional contexts the students might meet in the future. These draw directly on his KMT, but also his knowledge of real-life problems the content can solve (KoT – *phenomenology and applications*) and his knowledge of procedures (KoT – *procedures*); they draw, too, on his belief that mathematics serves a utilitarian purpose, and that by demonstrating this purpose to the students, their motivation will increase. In general, Carlos gives only one example to illustrate the procedure students should follow, and there is no conceptual emphasis or mention of potential errors.

Taking these results as typical cases of lecturers in linear algebra, we can identify the extent to which students were at the centre of the lecturer's thinking as a key differentiating element. A greater degree of concern for how students deal with the content could explain the richness of knowledge on display regarding how they learn (KFLM) and the importance placed upon how to deal with potential areas of difficulty, which can be connected to knowledge of strategies, tasks and examples for teaching (KMT). Within a context which seems to value an expository teaching style, this concern locates the teacher's beliefs more within a technological than a traditional model, in that the construction of content is promoted and there is greater interest in the students achieving a relational understanding (Skemp, 1976). In Carlos's case, a more extrinsic concern for content than for motivating the students was seen to be connected to an emphasis on an instrumental understanding, with less evidence of KFML and KMT in the teacher's practice.

With regard to previous studies, we note that, in line with the findings of Fukawa-Connelly et al. (2016), both lecturers' conception of teaching content favours exposition, albeit a mode of exposition incorporating a significant use of examples and, in Jordy's case, simulating the process of constructing content.

Our results seem to concur with those of Mamba et al. (2017) regarding the influence of teachers' content knowledge on their pedagogical content knowledge. In our study, Jordy's knowledge of the topic has a bearing on his awareness of potential difficulties and his choice of examples for teaching, while Carlos's knowledge – insofar as evidenced here limited to procedure – has a bearing (along with his phenomenological knowledge) only on the examples he selects.

We found no evidence of knowledge of the structure of mathematics (KSM) in our observations of the two lecturers, either in the episodes presented here or in the rest of the observation material across the whole study. KSM concerns knowledge of how items relate to each other, rather than knowledge of items in themselves. We wonder whether this kind of knowledge could be related to less traditional conceptions of mathematics teaching and learning (which would corroborate the results of Clark et al., 2014, in which, in certain cases, greater teacher knowledge was related to less traditional beliefs about the roles of student and teacher).

Our study, based on the class observation of two university lecturers in linear algebra, poses the hypothesis of possible nuclei of interconnected relations and beliefs which can account for key aspects of practice. The question of just how these conceptions and knowledge interconnect remains open. For example, Jordy's awareness of potential areas of difficulty would seem to be related to his conception of the role of error in teaching and learning. Some previous studies, such as the CGI programme (Carpenter & Fennema, 1992), fed teachers with information about what their students thought and noted subsequent adjustments by some teachers towards less traditional modes of teaching. What kind of knowledge about learning content might stimulate a modification in Jordy's conception of errors in the teaching and learning process?

The preliminary results of our study could constitute the point of departure for structuring the training and development of lecturers. An immediate example might be the consideration of the role of examples for helping students to overcome difficulties with linear algebra.

This study contributes data about university lecturers' knowledge and beliefs. In addition to accounting for what they do, it lays bare certain key aspects of their practice, such as the role of examples and errors, or the central issues which occupy their thoughts, such as the importance they confer on understanding content. In this way, we are confident that our work will continue to shed light on the practice and thinking of mathematics lecturers.

References

Ball, D. L., Thames, M. H., & Phelps, G. (2008). Content knowledge for teaching: What makes it special? *Journal of Teacher Education, 59*(5), 389–407.

Bardin, L. (1996). *Análisis de contenido*. Madrid: Ediciones Akal.

Biza, I., Giraldo, V., Hochmuth, R., Kharkbaz, A., & Rasmussen, C. (2016). Research on teaching and learning mathematics at tertiary level: State-of-the-art and looking ahead. In *Research on teaching and learning mathematics at the tertiary level*. Berlin: Springer International Publishing.

Britt, M. S., Irwin, K. C., & Ritchie, G. (2001). Professional conversations and professional growth. *Journal of Mathematics Teacher Education*, *4*(1), 29–53.

Carpenter, T. P., & Fennema, E. (1992). Cognitively guided instruction: Building on the knowledge of students and teachers. *International Journal of Educational Research*, *17*(5), 457–470.

Carrillo, J., Climent, N., Contreras, L. C., & Muñoz-Catalán, M. C. (2013). Determining specialised knowledge for mathematics teaching. In B. Ubuz, C. Haser, & M. A. Mariotti (Eds.), *Proceedings of the CERME8* (pp. 2985–2994). Antalya, Turkey: METU and ERME.

Carrillo, J., Climent, N., Montes, M., Contreras, L. C., Flores-Medrano, E., Escudero-Ávila, D. . . . Muñoz-Catalán, M. C. (2018). The Mathematics Teacher's Specialised Knowledge (MTSK) model. *Research in Mathematics Education*. ISSN: 1754–0178 (Online). doi:10.1080/14794802.2018.1479981

Carrillo, J., & Contreras, L. C. (1994). The relationship between the teacher's conceptions of mathematics and of mathematics teaching: A model using categories and descriptors for their analysis. In J. P. da Ponte & J. F. Matos (Eds.), *Proceedings of the 18th Conference of the International Group for the Psychology of Mathematics Education* (Vol. 2, pp. 152–159). Lisbon, Portugal: PME.

Carrillo, J., Montes, M. A., Contreras, L. C., & Climent, N. (2017). Les connaissances du professeurdansune perspective basée sur leur specialization: MTK. *Annuales de didactique et de sciences cognitives*, *22*, 185–205.

Charalambous, C. Y. (2015). Working at the intersection of teacher knowledge, teacher beliefs, and teaching practice: A multiple-case study. *Journal of Mathematics Teacher Education*, *18*, 427–445.

Clark, L. M., De Piper, J. N., Frank, T. J., Nishio, M., Campbell, P. F., Smith, T. M., . . . Choi, Y. (2014). Teacher characteristics associated with mathematics teachers' beliefs and awareness of their students' mathematical dispositions. *Journal for Research in Mathematics Education*, *45*(2), 246–284.

Dorier, J. L. (2002). Teaching linear algebra at university. In L. Tatsien (Ed.), *Proceedings of the International Congress of Mathematicians* (Vol. 3, pp. 875–884). Beijing, China: Higher Education Press.

Dunekacke, S., Jenßen, L., Eilerts, K., & Blömeke, S. (2016). Epistemological beliefs of prospective preschool teachers and their relation to knowledge, perception, and planning abilities in the field of mathematics: A process model. *ZDM Mathematics Education*, *48*(1), 125–137.

Ernest, P. (1991). *The philosophy of mathematics education*. London: The Falmer Press.

Even, R., & Ball, D. L. (2009). Setting the stage for the ICMI study on the professional education and development of teachers of mathematics. In R. Even & D. L. Ball (Eds.), *The professional education and development of teachers of mathematics: The 15th ICMI study* (pp. 1–9). Boston, MA, USA: Springer.

Fennema, E., & Franke, M. L. (1992). Teachers' knowledge and its impact. In D. A. Grouws (Ed.), *Handbook of research in mathematics teaching and learning* (pp. 147–164). New York: Macmillan.

Fernández, S., & Figueiras, L. (2014). Horizon content knowledge: Shaping MKT for a continuous mathematical education. *REDIMAT*, *3*(1), 7–29. http://10.4471/redimat.2014.38

Fives, H., & Buehl, M. M. (2012). Spring cleaning for the "messy" construct of teachers' beliefs: What are they? Which have been examined? What can they tell us? In K. R.

Harris, S. Graham, & T. Urdan (Eds.), *APA educational psychology handbook: Individual differences and cultural and contextual factors* (Vol. 2, pp. 471–499). Washington, DC, US: American Psychological Association. http://dx.doi.org/10.1037/13274-000

Flores-Medrano, E., Escudero-Ávila, D., & Carrillo, J. (2013). A theoretical review of specialized content knowledge. In B. Ubuz, C. Haser, & M. A. Mariotti (Eds.), *Proceedings of the CERME8* (pp. 3055–3064). Antalya, Turkey: METU & ERME.

Flyvbjerg, B. (2011). Case study. In N. K. Denzin & Y. S. Lincoln (Eds.), *The Sage handbook of qualitative research* (pp. 301–316). Thousand Oaks, CA: Sage Publications.

Forgasz, H., & Leder, G. (2008). Beliefs about mathematics and mathematics teaching. In P. Sullivan & T. Wood (Eds.), *The international handbook of mathematics teacher education: Knowledge and beliefs in mathematics teaching and teaching development* (pp. 173–192). Rotterdam, The Netherlands: Sense Publishers.

Fukawa-Connelly, T., Johnson, E., & Keller, R. (2016). Can math education research improve the teaching of abstract algebra? *Notices of the AMS, 63*(3), 276–281.

Hernandes-Gomes, G., & González-Martín, A. S. (2015). Mathematics in engineering: The professors' vision. In K. Krainer & N. Vondrová (Eds.), *Proceedings of the CERME9* (pp. 2110–2116). Prague, Czech Republic: ERME.

Johnson, E., & Larsen, S. P. (2012). Teacher listening: The role of knowledge of content and students. *Journal of Mathematical Behavior, 31*, 117–129.

Kuhs, T. M., & Ball, D. L. (1986). *Approaches to mathematics: Mapping the domains of knowledge, skills and dispositions*. East Lancing: Michigan State University, Center on Teacher Education.

Leatham, K. R. (2006). Viewing mathematics teachers' beliefs as sensible systems. *Journal of Mathematics Teacher Education, 9*(1), 91–102.

Lin, F.-L., & Rowland, T. (2016). Pre-service and in-service mathematics teachers' knowledge and professional development. In A. Gutierrez, G. C. Leder, & P. Boero (Eds.), *The second handbook of research on the psychology of mathematics education* (pp. 483–520). Rotterdam, The Netherlands: Sense Publishers.

Mamba, F., Mosvold, R., & Bjuland, R. (2017). A preservice secondary school teacher's pedagogical content knowledge for teaching algebra. In T. Dooley & G. Gueudet (Eds.), *Proceedings of the CERME10* (pp. 3336–3343). Dublin, Ireland: ERME.

McCrory, R., Floden, R., Ferrini-Mundy, J., Reckase, M. D., & Senk, S. L. (2012). Knowledge of algebra for teaching: A framework of knowledge and practices. *Journal for Research in Mathematics Education, 43*(5), 584–615.

Neuman, W. L. (2014). *Social research methods: Qualitative and quantitative approaches*. Harlow: Pearson Education Limited.

Pajares, M. F. (1992). Teachers' beliefs and educational research: Cleaning up a messy construct. *Review of Educational Research, 62*(3), 307–332. doi:10.3102/00346543062003307

Schoenfeld, A. H. (2010). *How we think*. New York, NY: Routledge.

Schoenfeld, A. H., & Kilpatrick, J. (2008). Toward a theory proficiency in teaching mathematics. In T. Wood & D. Tirosh (Eds.), *Tools and processes in mathematics teacher education* (pp. 321–354). London: Sense Publishers.

Shulman, L. S. (1986). Those who understand: Knowledge growth in teaching. *Educational Researcher, 15*(2), 4–14.

Shulman, L. S. (1987). Knowledge and teaching: Foundations of the new reform. *Harvard Education Review, 57*(1), 1–22.

Skemp, R. (1976). Relational understanding and instrumental understanding. *Mathematics Teaching, 77*, 20–26.

Stake, R. E. (2003). Case studies. In N. Denzin & Y. Lincoln (Eds.), *Strategies of qualitative inquiry*. Thousand Oaks, CA: Sage Publications.

Thompson, A. (1992). Teachers' beliefs and conceptions: A synthesis of the research. In D. A. Grouws (Ed.), *Handbook on mathematics teaching and learning* (pp. 127–146). New York: Macmillan.

Törner, G. (2002). Mathematical beliefs: A search of a common ground: Some theoretical considerations on structuring beliefs, some research questions, and some phenomenological observations. In G. C. Leder, E. Pehkonen, & G. Törner (Eds.), *Beliefs: A hidden variable in mathematics education?* (pp. 127–147). Dordrecht: Kluwer Academic Publishers.

Vasco, D., & Climent, N. (2016, October 5–7). Relationships between the knowledge and beliefs about mathematics teaching and learning of two university lecturers in linear algebra. In S. Zehetmeier, B. Rösken-Winter, D. Potari, & M. Ribeiro (Eds.), *Proceedings of the Third ERME Topic Conference on Mathematics Teaching, Resources and Teacher Professional Development* (pp. 177–186). ETC3. Berlin, Germany: Humboldt-Universitätzu Berlin. ISBN: 978-3-00-058755-9.

Vasco, D., Climent, N., Escudero-Ávila, D., Montes, M. A., & Ribeiro, M. (2016). Conocimiento Especializado de un Profesor de Álgebra Lineal y Espacios de Trabajo Matemático. *Bolema, 30*(54), 222–239. http://dx.doi.org/10.1590/1980-4415v30n54a11

Watson, A., & Mason, J. (2005). *Mathematics as a constructive activity: Learners generating examples.* Mahwah, NJ: Lawrence Erlbaum Associates.

Wilson, M., & Cooney, T. (2002). Mathematics teacher change and development. In G. C. Leder, E. Pehkonen, & G. Törner (Eds.), *Beliefs: A hidden variable in mathematics education?* (pp. 127–147). Dordrecht: Kluwer Academic Publishers.

Zaslavsky, O. (2010). The explanatory power of examples in mathematics: Challenges for teaching. In M. K. Stein & L. Kucan (Eds.), *Instructional explanations in the discipline* (pp. 107–128). Boston, MA, USA: Springer.

8

USING THE *DISCOURSES OF LEARNING IN EDUCATION* MAPPING TO ANALYSE RESEARCH INTO MATHEMATICS TEACHER EDUCATION AND PROFESSIONAL DEVELOPMENT

Laurinda Brown & Brent Davis

1. Introduction

At CERME5 in 2007, in what was then Working Group 12 (WG12), *From a study of teaching practices to issues in teacher education*, José Carrillo organized a session where he invited, before the conference, a group of participants to use their differing models of analysis, introduced in their papers, to interpret a section of videotape. The methods and results of their analyses were presented during Panel 3 and included Rowland's (local to mathematics education) knowledge quartet, psychoanalytic theory and Cole and Engeström's (widely used in research into social practices) activity theory, developed from the work of Vygotsky and Leont'ev. What was surprising to Laurinda, as one of the presenters in Panel 3, and to the members of WG12, was that these different tools for analysis pointed to the same sentence for further discussion, an articulation by the teacher in the videotape (Carrillo, Santos, Bills, & Marchive, 2007, p. 1824).

This experience came to Laurinda's mind when considering the suggested focus for this chapter, *Theoretical/analytical framework for analysing mathematics teacher education and professional development*, with its singular framework. There is a proliferation of frameworks in use, but, given the experience of different analyses pointing to the same event, one question posed in CERME5, WG12, was, "Is it possible to speak about a meta-model?" (Carrillo et al., 2007, p. 1824). The frameworks that are chosen by researchers depend on their experiences and contexts; they are not arbitrary. Over time, in CERME thematic working groups related to mathematics teacher education and professional development, many frameworks reappear.

At a meta-level, however, learning is at the centre of research related to mathematics teacher education and the professional development of teachers of mathematics; for instance, the learning of mathematics teacher educators in self-study about the development of their own practices; the learning of teachers of mathematics developing their teaching practices; and the learning of children in mathematics classrooms engaging in learning mathematics. The research is about learning, but what is learning? Oriented by the realization that there are literally hundreds of theories of learning at play in contemporary education, we offer an analysis of theoretical perspectives underpinning the methodological aspects of research in the field, with the aim of providing a tool to support researchers in designing their own research and in analysing the research of others. To assist in this project, our writing is framed with a map, a meta-model, perhaps, of *Discourses on Learning in Education* (https://learningdiscourses.com/). After a little background on how this chapter came to be written, we introduce the organizational details of the meta-model of learning discourses. We then use the meta-model to analyse research contributions to CERME10 (Dooley & Gueudet, 2017), the most recent ERME conference with available proceedings at the time of writing, using discourses in two example cases, Variation Theory and Professional Learning Communities, raising issues related to the analysis of mathematics teacher education and development to point to ways forward, strengths and areas for development in the field.

2. Background

How did this chapter come to be written? We, the authors, have worked together on a number of writing and editing projects, including pulling together one of the chapters for the ICMI Study 15 volume, *The professional education and development of teachers of mathematics* (Davis & Brown, 2009). We take the following, from that chapter's concluding remarks, as underpinning and motivating discussions in this chapter:

> What is being said . . . recognizes the complexity of teacher development. Knowledge is situated in the contexts of the practices of countries, within both mathematics learning and teaching practices. An individual teacher's practice and development is difficult to talk about in isolation without considering the local nature of the practices involved at different focal lengths of the lens viewing those practices, ranging from governmental, school, particular class culture, and achievements of students. To be able to consider working with teachers who are themselves learning about the teaching and learning of mathematics [often] working together in groups . . . seems like an important construct. Co-observation and co-teaching – being able to act in different ways through observing a different reality, planning together, or developing a culture within a department or within a classroom that itself becomes a learning community or [learning] classroom . . . gives us a way of thinking about professional development where the teachers share thoughts and practices rather than a particular way of doing things. Teachers learn, and those who teach teachers learn correspondingly.
>
> (p. 165)

More than 10 years later, as this chapter was written, we have continued to be immersed in working with teachers, often in collaborative groups. We take the closing remarks above to apply to all the writing in this chapter, motivating especially our focuses on learning, and professional learning communities within mathematics teacher education and professional development, an area that has developed strongly since 2009. We now introduce the organization of the meta-model map, *Discourses on Learning in Education*, aiming to illustrate how it might be useful for thinking about and across the many theoretical perspectives about mathematics learning that have been or might be brought to bear on mathematics teacher education.

3. Variation among discourses on learning in education

What is learning? How does it happen? Can it be made to happen?

It turns out that these questions have been answered in many, many ways. Davis developed the learningdiscourses.com website in an effort to present an accessible survey of some of those responses. The site includes summaries of more than 600 discourses as it provides information on their foci, themes, imagery, related discourses and supporting evidence. Brief genealogical details are also offered, especially for those discourses with diverse interpretations and/or multiple sub-discourses. This writing is intended to be quasi-interactive. It is suggested that, as you read on, you open the mapping and explore as we describe how we worked.

The heart of learningdiscourses.com is a mapping of labels of discourses of learning. To explore the mapping, imagine that the placements of the labels are within the first quadrant of a Cartesian graph, with vertical and horizontal axes having particular meanings. The mapping makes use of several devices to highlight convergences and divergences among discourses.

These devices include the following four, which will be listed first before being discussed:

- Firstly, clustering of thematically similar discourses, and clustering some of those clusters into meta-clusters. As you hover over the label of a meta-cluster, it turns blue and the labels in the meta-cluster are shown within a highlighted area of the map.
- Secondly, using the vertical axis to draw distinctions between discourses concerned mainly with theorizing learning and those concerned mainly with influencing learning.
- Thirdly, using the horizontal axis to organize discourses according to core metaphors (e.g. object-based theories vs. agent-based theories), associated imagery (e.g. tendency toward points, lines, discrete regions and Euclidean shapes vs. preference for fractal-like, nested and networked forms) and key conceptual commitments (e.g. embracing or rejection of such dyads as objective/subjective, internal/external, mind/body, self/other, individual/collective and human/nature).

- Fourthly, using colour-coding to identify the "scientific" status of each discourse – where the notion of *scientific* is operationalized in terms of the extent to which a discourse is explicit about its assumptions, the existence of a body of replicable supporting evidence, the absence of opposing evidence, the availability of the discourse for revision and the existence of complementary discourses.

Clicking on any name or label on the map triggers a pop-up window with a concise description of the discourse(s) associated with that tab as well as a "Learn More" link that takes users to a file with additional information, including critiques, influential figures and grounding metaphors. These files are not offered as comprehensive or authoritative summaries of individual discourses; rather, the principle aim is to offer users glimpses into convergences and divergences among discourses on learning that are currently at play.

The following discussion will show how we came to choose the labels we will use in the following two exemplar cases, which will illustrate the use of the mapping as a tool in analysing research into mathematics teacher education and professional development. The process of choosing the labels also illustrates the use of the meta-model in supporting the design of research.

Firstly, we looked beneath the meta-cluster labels. It is possible to click on both the labels for the meta-clusters and labels for clustered items. For instance, in searching for the meta-cluster that most closely fits the image of professional development described in the ICMI study quotation, above, we moved away from those that focus on aspects such as the individual learner or that embrace the separation of mind and body. This took us towards the right-hand side of the horizontal axis, where the key metaphor for learners was interactive participants, which fits well with our focus on collaboration with teachers. The key metaphors for learning around this area of the mapping are stated as participating and expanding the possible. We did not want a deficit model of teachers' practices. Our image, as stated in the 2009 concluding remarks above, was that in working together, exploring each other's practices, the teachers and the mathematics teacher educator(s) forming the collaborative group are all becoming able to act in different ways to their habitual practices. Using the language of the mapping, the learning here is not about acquiring, attaining or mastering, but about co-observation, co-learning and expanding the space of the possible.

We searched vertically around this position on the horizontal axis, exploring each meta-cluster by clicking to find its synopsis. The synopsis for the meta-cluster socio-cultural-focused discourses was a good fit for mathematics teacher educators working with groups of teachers or/and prospective teachers collaboratively. According to the synopsis, socio-cultural discourses:

> tend to operate from the assumption that collective knowing unfolds from and is enfolded in individual knowers. Consequently, most of these discourses attend [to] the situated learner and/or the collective learning

system – rather than the individual learner. Matters that figure promi-
nently include context, participation, collaboration, ethics, democratic
obligation, and tacit norms.

(Davis & Francis, 2019a)

Having decided to focus on socio-cultural-focused discourses, the area on the
mapping spreads over all of the vertical spectrum, so it seemed unlikely that,
in this chapter, we would be able to discuss all the labels in this meta-cluster.
The lower-half of the vertical axis points to discourses that are concerned with
theorizing and the upper-half with influencing learning. How were we to limit
our focus? Within the colour-coding of labels, green indicates the most scientific, described in a key on the map as being "robustly theorized and empirically
grounded, aimed at innovative contributions to understanding learning". We
were interested in exploring green labels.

Our initial strategy was to focus on the subset of contemporary discourses on
learning in education that are principally concerned with how collectives unfold
from and are enfolded in individuals. We thus homed in on the following inter-
secting clusters, all within the meta-cluster of socio-cultural-focused discourses:

- Collectivist learning theories, which are concerned with the emergence and
  maintenance of both individual knowing and collective knowledge, recognizing these dynamic phenomena to be inextricably intertwined and continuously co-emergent
- Discourses on learning collectives, which are concerned with matters of
  designing tasks, designating roles and structuring situations in ways that support the maintenance and elaboration of teams or organizations
- Language-focused discourses, which attend to the role of symbol systems
  in constituting and maintaining knowers' realities, as well as to their role in
  enabling and constraining personal possibilities within those realities
- Ecological discourses, which typically foreground multiple forms of relationship (e.g. biological, social, cultural, environmental, epistemological)

Each of these comprises 10 to 20 specific theories.

It was only as we reviewed papers from CERME10 that we started to be aware
of the considerable breadth associated within socio-cultural focused discourses,
spanning such matters as the situated learner, collective learning systems, cultural dimensions of knowing and knowledge, social contracts, scripts, position-
ings, context, participation, collaboration, ethics, democratic obligation, tacit
norms and critical awareness. With this more detailed knowledge, we elected to
revise our tactic, focusing our efforts through two entries on the map: Variation
Theory and Professional Learning Community to act as exemplars of using the
mapping as an analytical tool.

Our choice of Variation Theory, within ecological discourses, within socio-
cultural focused discourses, labelled green, was motivated by two main factors.

Firstly, reviewing CERME10 papers, there is considerable uptake of Variation Theory within mathematics teacher education and professional development. That uptake also demonstrates considerable variation! Secondly, the discourse is located somewhat ambiguously on the mapping. It has been placed in the region between those discourses concerned with theorizing learning and those more concerned with influencing learning – or, in more familiar terms, it is situated as both a theory of learning and a theory of teaching and is nudged more toward the latter on the map.

The choice of Professional Learning Community, among "discourses on learning collectives" and within "socio-cultural focused discourses", was prompted by our focus on collaborative learning. It is nested among "discourses that attend simultaneously to individual and collective learning", within the upper half of the vertical axis labelled influencing learning. We considered that the combination of analyses, using the mapping, into the uptakes of Variation Theory and Professional Learning Community, within CERME10, would afford an opportunity to examine how discourses are adopted and how they are adapted in the field of mathematics education related to mathematics teacher education and professional development.

4. First example case: Variation Theory

Clicking on the label for Variation Theory brings up the synopsis:

> Variation Theory draws together insights into human attention and the structure of different knowledge domains to offer advice on critical discernments that are necessary to a discipline, strategies to channel learners' attentions to those discernments, and tactics to encourage meaning making through the juxtaposition of those discernments.
>
> *(Davis & Francis, 2019b)*

To make explicit the perspective for analysis, there is a button called *Learn more* at the foot of the synopsis that can be clicked to bring up more information (included in full here; the items in bold are other labels that can be accessed through the mapping). The headings for this further information are the same for all the discourses on learning, providing a tool for analysis of any theories:

Focus

Exploiting habits of perception to support conceptual learning

Principal metaphors

- Knowledge is . . . evolving webs of coherent interpretations
- Knowing is . . . acting and interpreting (based on one's history)
- Learner is . . . a noticer and integrator (individual)

- Learning is . . . construing, connecting, interpreting, weaving
- Teaching is . . . challenging attentions and juxtaposing critical features

Originated

1990s

Synopsis

Variation Theory is based on four key principles from **Cognitive Science**, namely that (1) the capacity of working memory is very limited, (2) every experience presents innumerable features that might consume one's attention, (3) established collective knowledge tends to be well structured and reliant on highly specific "critical features," and (4) humans are exquisitely attuned to change/difference. **Variation Theory** thus offers focused advice on strategies for channeling learners' limited attentions to critical features (e.g., systematic variation of the critical feature while holding all other features constant) in supporting robust conceptual development.

Commentary

Relative to other educational theories, **Variation Theory** has a very narrow focus. For that reason, it tends to assume rather than assert key principles of learning – for example, the tenet that what one learns is principally conditioned by both what one already knows and what one is currently noticing. (Other examples are shared with various **Embodiment Discourses**.) **Variation Theory**'s advice on channeling learner attentions is theoretically sound and empirically grounded. However, that advice is only useful if accompanied by rather sophisticated knowledge of the structures of concepts under study. It's one thing to prompt awareness to critical features, and it's quite another to do so in sequences that support sound construals.

Authors and/or prominent influences

Ference Marton

Status as a theory of learning

Proponents of **Variation Theory** often assert that it is both a theory of learning and a theory of teaching, given that it is grounded in principles of **Cognitive Science** and that it offers pragmatic advice for educators. Because it does not develop or extend insights into the complex dynamics of learning, we do not classify it here as a theory of learning.

Status as a theory of teaching

In our analysis, **Variation Theory** is principally a theory of teaching.

Status as a scientific theory

Variation Theory is gaining a strong foothold in some areas of educational research, particularly those associated with the STEM domains. Emerging results, coupled to its solid theoretical grounding, appear sufficient to assert that **Variation Theory** is a scientific theory.

(Davis & Francis, 2019b)

If this meta-model can be used as a tool to analyse the research papers produced for CERME10 (Dooley & Gueudet, 2017), then it will let us see something new that we would not have been aware of without it. What we observe will then point to issues and ways forward in future research. So, we will now analyse the use of Variation Theory in papers given at CERME10.

4.1 Citations and references

In the information provided by the mapping, Ference Marton is mentioned as author. Is it the case that each of the papers using Variation Theory refers back to sources such as Marton and Booth's (1997) book *Learning and awareness*, or perhaps directly to Marton's (1970) thesis or later research projects, the progression in ideas of which is documented in the 1997 book? The process of identifying the citations to the work of Marton in the papers from CERME10 illustrated Variation Theory in that the papers that did use Marton's original work (such as Odindo, 2017, pp. 3129–3136; Runnesson Kempe, Lövström & Hellqvist, 2017, pp. 3153–3160; and Tirosh, Tsamir, Barkai & Levenson, 2017, pp. 1917–1924) did so using different references (Marton references are cited in order of use in the respective papers, with two papers cited in the first source, Marton, 2015; Marton & Booth, 1997; Marton & Pang, 2003; Ling & Marton, 2012).

Not all papers using Variation Theory go back to Marton's original sources. Since CERME papers have an eight-page limit, this may be, for instance, because space is tight. Watson and Mason (2005), from the UK research community, who do cite Marton's work, have extended Variation Theory to *dimensions of possible variation* (what is possible to vary) and *range of permissible change* (perceived constraints on the extent and nature of change in any of the dimensions of variation). Mason developed the discipline of noticing (2002), so, given one metaphor for Variation Theory is of the individual learner as noticer, his interest arising in this theory is clear, from a socio-cultural-focused perspective, linked to his history. Some researchers go directly to Watson and Mason without citing Marton; for instance, Venkat and Askew (2017) appeal to Watson and Mason's (2005) example spaces, through employing a "range of permissible variation in the example space" (2017, p. 3197). The authors (Sakonidis, Drageset, Mosvold, Skott & Didem Taylan, 2017) of the introduction to the set of papers in thematic working group 19 attribute Venkat and Askew's writing to Variation Theory, saying, "Venkat and Askew employ variation theory and example spaces to understand how teachers mediate primary mathematics, mainly how they generate and validate solutions as well as build mathematical connections" (p. 3037).

The mapping focused our attention, making us aware of use of primary sources as opposed to secondary sources by authors.

4.2 Contextual factors

Variation Theory is used commonly in Scandinavian countries, Marton's home area. This is perhaps not surprising, as Marton has written with such researchers as Runesson, who has also written extensively and independently about the theory (for instance, Runesson, 2005). Research communities grow up that attract research students. Combining Variation Theory with another perspective seems common, such as with lesson study (Dooley & Aysel, 2017, pp. 3758–3760), the combination becoming learning study, a "theory-informed version of [l]esson study" (Marton & Pang, 2003). Developments in Scandinavia to the original theory give evidence for the greenly labelled scientific nature of the theory. Runnesson Kempe et al. (2017) use learning study, saying that it shares with lesson study "collaboration among teachers and the iterative design of planning, implementing, observing and revising of the lesson, but it is framed by a theory of learning – variation theory" (p. 3154). Björklund (2017), from the University of Gothenburg, Sweden, where Marton studied for his doctorate, focuses on Marton's work with early number that includes

> a bold conjecture that the understanding of whole numbers originates from experiences of different aspects of number, rather than a predictable development trajectory. Focus is here shifted from descriptions of children's competences and learning trajectories towards the content to be learnt, and particularly what it takes to learn that content.
>
> *(p. 1821)*

This focus, although in the paper describing Variation Theory as a theory of learning, gives support to the mapping's placing of variation as a theory of teaching, "what it takes to learn that content".

Variation Theory has been taken up outside of Scandinavia. For instance, the phrase is used in the UK as part of the government's push, through the network of Maths Hubs across the country, for Mastery Teaching and Learning (National Centre for Excellence in the Teaching of Mathematics – NCETM – www.ncetm.org.uk). This has led to mathematics teacher educators having a research focus, such as Baldry (2017) at the University of Leicester, looking at *A teacher's orchestration of mathematics in a "typical" classroom* (pp. 3041–3048), aiming to see what current practices of teaching look like. In this paper, a range of theoretical perspectives, including Variation Theory (Marton & Pang, 2006), were used to develop "[a]n observation framework . . . to interpret classroom activities" (p. 3041). Watson (2018), from the UK, has influenced the spread of Variation

Theory, including through collecting together a series of articles from the journal of the Association of Teacher of Mathematics (ATM), *Mathematics Teaching*, as a resource book (available through the ATM website) for teachers.

One question arising for us here is related to the contexts, such as cultural and geographical, in which a theory arises and its spread. Is the spread to contexts with similar mindsets and/or problems? A mapping of where discourses of learning are used geographically, and their spread over time, would be an interesting extension of the mapping tool.

4.3 Theory of learning and/or theory of teaching

The mapping tool places Variation Theory as a theory of teaching, not of learning. It is sometimes difficult to untangle these two categories of emphasis. Some of the papers describe the "Variation Theory of learning" without specifying what is meant by *learning*; others say simply "Variation Theory". We have not yet found a paper that describes "Variation Theory of teaching". Variation Theory relates to the different ways of experiencing some event. For instance, in his 1997 book, Marton (1997) describes five different ways of experiencing number, number facts, finger numbers, counted numbers, numbers as extents and numbers as names, developed in a phenomenographic study by Neumann. From this theoretical perspective, learning is defined as "a change in the way something is seen, experienced or understood" (Runesson, 2005, p. 70). Thus, small changes in the design of a task can result in changes in what it is that students discern or notice. The aim is not to experience the ways of experiencing number as discrete but to be able to develop number concepts through experiencing more than one aspect at any time. Although the focus seems to be on experiencing learning, in practice, as seen in Björklund's (2017) paper referring to the "variation theory of learning", the focus on what it takes to learn content implies a facilitating other directing their own attention to what aspects of numbers are discerned by toddlers in play and interaction in pre-school. Who is making the changes in the design of a task? This is one definition of teaching, not about transmission, that supports the view of Variation Theory influencing teaching.

Odindo (2017) also describes the "Variation Theory of learning". However, again – and here we will give a more extended quotation from the paper – it is difficult to disentangle the teaching and learning aspects; they seem to arise together. The section begins with "Variation Theory is a theory of learning which asserts that to learn something entails experiencing it in a variety of ways" (p. 3130). There is then a focus on achieving this: "teachers need to create learning opportunities by explicitly or implicitly offering patterns of variation in which some parts remain invariant as others vary" (p. 3130), followed by an extended and useful discussion of an example related to quadrilaterals and different forms of variation.

The point here, though, is that, in this theory, learning and teaching are coupled by interpreting learning in terms of *noticing what has not yet been noticed*, rather than *knowing what is not yet known*. Teaching, then, comes to be about channelling learners' attentions and juxtaposing acts of noticing. Consequently, the teacher's attention is on the learner's immediate perceptions, not on some pre-given learning trajectory. That is, the teacher is more concerned with sequences of strategies to channel noticing and sequences of discernments, not with clear explanations and sequences of steps. Hence, the placement of Variation Theory on the mapping centrally between those discourses principally concerned with theorizing learning and discourses principally concerned with influencing learning (i.e. teaching) seems crucial to working with Variation Theory. The discourse is certainly informed by well-grounded principles of cognition, but its main focus is on applying those principles rather than on developing or extending them to better understand the complex phenomenon of learning.

This same distribution of foci can be found in the questions posed by Tirosh et al. (2017), where the researchers use Variation Theory of learning with preschool teachers. Their first question was, "Given an extension task and a set of repeating patterns, what are the various ways preschool teachers implement the task?" (p. 1918), followed by, "What can we learn about children's patterning abilities from the different implementations?" (p. 1918) The second question arises from the first considering that "learners may experience objects in various ways". In the mix here is that the observer/researcher focused on the experiences of, firstly, the preschool teachers and then the children in what seems to be an elegant design. However, here the preschool teachers' differing implementations and the children's learning are both foregrounded, with the main interest being influencing learning rather than theorizing learning.

In fact, to our analysis, despite the much stronger tendency to attach Variation Theory to learning rather than to teaching, every one of the CERME10 papers that we analysed was much more concerned with influencing learning rather than theorizing learning. That observation comes with the caveat that all of these papers appear to be anchored to the tacit conviction that learning is always dependent on, but never determined by, teaching. Thus, while none offers significant new insights into what learning is and how it happens, all tether learning to teaching in complex ways.

That realization signals a vital link to our second case, on more collective-focused discourses on learning. Within Variation Theory, the learner is consistently framed and described as an individual. However, learners are understood to be co-influencers within systems of learners, collectively entangled in exquisite dances of noticing and associating. While the learner is seen an individual within Variation Theory, there is a strong sense that an individual is not an isolated entity; the attentional systems of individuals are always and already dynamically coupled. Phrased differently, while it is not an explicit tenet of the discourse, Variation Theory is articulated in a manner that invites an awareness of how a collective of learners can be understood as a coherent cognitive entity.

Other discourses take that point as the explicit focus rather than an implicit assumption, as is exemplified in our second case.

5. Second example case: Professional Learning Community (PLC) and collaborative learning

Clicking on the label for "Professional Learning Community" brings up the synopsis:

> While conceptions and definitions vary considerably, most often a Professional Learning Community (PLC) is understood as an approach to collaborative learning among professional colleagues in specific work environments. PLC is among those methods that regard the collective as a learning system – that is, not just a collection of individual learners or a context to support individual learning, but a collective learner.
>
> *(Davis & Francis, 2019c)*

Pressing the button *Learn more* brings attention to the following information, given in full, that will be used to analyse CERME10 papers similarly to how Variation Theory was discussed in the first case study:

Focus

Effective community action in a professional setting

Principal metaphors

- Knowledge is . . . sum of expertise in the community
- Knowing is . . . effective collective action
- Learner is . . . a member (individual) or a professional project (collective)
- Learning is . . . simultaneous and integrated modification of individual habits and collective process
- Teaching is . . . co-teaching – that is working collaboratively to positively affect one another's habits of acting

Originated

1990s

Synopsis

While conceptions and definitions vary considerably, most often a **Professional Learning Community** (**PLC**) is understood as an approach to **Collaborative Learning** among professional colleagues in specific work environments. Embracing **Socio-Cultural Theory** and related discourses, **PLC** is among those methods that regard the collective as a learning system – that is, not just a collection of individual learners or a context to support individual learning, but a collective learner aiming to transform its own culture/character (contrast **Learning Community**).

An effective **PLC** entails shared commitments to self-critique, collaboration, and reflection around matters of professional purpose.

Commentary

Most criticisms of **PLC** are focused on trivialized versions or ineffective implementations. Regarding the former, versions of **PLC** that are attentive to their conceptual roots in **Socio-Cultural Theory** tend to escape extensive criticism. Regarding the latter, the effectiveness of any practical approach to change – especially one that purports to simultaneously support individual and collective transformation – are limited by the understandings, intentions, and commitments of those involved. In a broader culture of competitive individualism, it's not hard to imagine potholes for an approach that focuses on collaboration and collectivity.

Authors and/or prominent influences

Peter Senge
Shirley M. Hord

Status as a theory of learning

PLC is not a theory of learning.

Status as a theory of teaching

While not a theory or model of teaching, it is not a stretch to suggest that **PLC** is a model of co-teaching – that is, of working together to positively affect one another's habits of acting and interpretation, aiming to improve the ethos and effectiveness of a professional community. To that end, **PLC** offers extensive advice on identifying shared goals, articulating visions, seeking solutions, working collaboratively, conducting inquiries, sharing accountability, encouraging experimentation, questioning habits, and ensuring critical reflection.

Status as a scientific theory

The most prominent versions of **PLC** are informed by and attentive to **Socio-Cultural Theory** and related perspectives. While **PLC** is supported by targeted empirical research, it appears that the bulk of its support is derived from its associated scientific theories of learning.

(Davis & Francis, 2019c)

As with our first case study, the items in bold are labels on the mapping, so information can be found from the website. The process of analysis of PLC research in CERME10 papers proved quite different from the first case study.

5.1 Citations and references

A search was undertaken within the CERME10 proceedings for papers using PLC and also for the authors Senge and Hord. Senge had no hits and Hord only one. Professional Learning Community had only one hit. We had known a lot of research was being done using Variation Theory, but the almost non-existence of hits for PLC was surprising for us. Given our quotation presented earlier, we had wanted to move beyond collaboration – which is now widely accepted, warranting, for instance, an ICME international survey (Robutti et al., 2016) and an ICMI study conference, *Teachers of mathematics working and learning in collaborative groups* (February 3–7, 2020) – to the insights that can be drawn through considering the community. What is happening?

To confirm to ourselves that the work on PLC, labelled green, did have scientific standing in the mathematics education and education communities, we looked at two recent sources reviewing collaboration in learning (Slavit, 2020; Hargreaves, 2019). In Volume 3 of the recently produced second edition of the *International handbook of mathematics teacher education*, Slavit (2020) has written a chapter on frameworks for analysing collaborative teacher activity. His approach was to focus on collaboration amongst classroom teachers. On pages 25–26 of Slavit's chapter, Hord's (1997) work on Professional Learning Community is referred to. In this research, which she undertook after experiencing a learning community using Senge's ideas, she "identified five dimensions of PLCs critical to effective learning communities: (1) shared values and vision; (2) shared and supportive leadership; (3) collective learning and application; (4) shared personal practice and (5) supportive conditions" (Anfara & Mertz, 2015, p. 224), whilst "exploring learning communities in school settings from a conceptual standpoint" (p. 224). Here is a framework that can be used to analyse collaborative groups to distinguish between, say, Professional Learning Communities and any group of educators that meet together for any purpose. In Hargreaves's (2019) personal exploration of 30 years of researching collaboration, the discussion related to PLCs is a substantial part of the paper. He builds up to his current use of the phrase *collaborative professionalism* as an extension of PLC to allow for a move away from the extremes, when PLCs do not thrive, of the community being used for the implementation of top-down initiatives on the one hand and becoming uncritical tea-sharers whose communication norm is politeness on the other. This move is "not just *between* practices that are too comfortable or contrived, but somehow above and *beyond* these polar opposites" (p. 612). Where PLCs were successful, "professional learning communities were not just a collection of meetings but a way of life" (p. 611). With the development of PLC to *collaborative professionalism*, PLC does have grounds to be named a scientific theory. Given that the wider educational community to CERME finds PLC to be a powerful learning discourse, what is happening in mathematics education?

Within mathematics education there was an ICMI International Survey on teachers working and learning through collaboration (Robutti et al., 2016).

This paper reports being based on an earlier ICMI survey in 2004 (Adler, Ball, Krainer, Lin & Novotna, 2005) that was focused on the then "emerging field" of mathematics teacher education, not specifically collaboration. There is evidence here of how the research in the field has moved on. In the 2016 survey, references to Professional Learning Communities exist in a paragraph entitled "Other methodological approaches" (p. 677): "We found eight papers that referred to *Professional Learning Communities* or PLC. In all of these, the communities described and their activity overlap in differing ways with areas of community" (Robutti et al., 2016, p. 677, emphasis in the original). There is no reference to the work of Senge or Hord; in fact, there seems to be no one theoretical formulation of PLC linking these eight papers. The most common theories used in this survey in relation to community were reported (p. 669) as Community of Practice, Valsiner's Zone Theory of Child Development and, largely developing out of the work of Vygotsky and Leont'ev, Activity Theory. Community is ascribed to Engeström's (1999) Activity Theory and Lave and Wenger's (1991) Community of Practice. The socio-cultural perspective is accepted for these two theories, and they both appear in the Socio-Cultural Focused area of the mapping. Cultural-Historical Activity Theory is labelled green and Community of Practice brown, indicating that it "lacks some critical element associated with robust, scientific knowledge" according to the website. Valsiner's Zone Theory of Child Development appears labelled red, signifying a theory being inattentive to assumptions, placed in "Unaffiliated Discourses". In what follows, we will concentrate briefly on the two theories exploring aspects of community within the CERME10 papers.

It is not that different papers are used by different authors to support their work or that secondary sources are quoted instead of those of the originator. Here, the process has uncovered an important but missing discourse on learning that could allow greater awareness by researchers and mathematics educators when working collaboratively, both as mathematics teachers with children learning mathematics and as mathematics teacher educators working with teachers or prospective teachers.

5.2 Contextual factors

Given that searching on Professional Learning Community pointed to a gap in current research in mathematics education at CERME10, contextual factors will be discussed through a search related to community to gain a sense of the prevalence of the two theories. We considered searching on collaboration/collaborative, but this generated a wealth of papers without, necessarily, more developed theory underpinning them.

It is clear that learning in collaborative groups of teachers happens, for instance, in Lesson Learning using Variation Theory. The Runnesson Kempe et al. (2017) paper is the one that had the only hit on Professional Learning Community, and that was in the title of one of the references in the paper.

There are also communities at the national level such as IREMs in France, comprising a mix of roles in the community, such as, teachers and university teacher educators (Llinares, Krainer & Brown, 2018). Jaworski (2003) has led research/curriculum development projects in Norway, including educators such as Goodchild and Fuglestad, with teachers in a collaborative group. Beneath the general educational heading of Collaborative Learning, there are a number of local to mathematics education terms. Searching the CERME10 proceedings for Jaworski finds the following in Nardi's plenary address, "I note that 'co-learning partnership' is a term that I had become familiar with from the work of my doctoral supervisor and research collaborator Barbara Jaworski (2003)" (Nardi, 2017, p. 14). Work is being done in the area of PLCs, but the language used (e.g. Co-learning Partnership, Jaworski, 2003, 2005) is not the same throughout the field.

Here are pointers to a few of the papers in CERME10 related to the ideas of Community of Practice and Activity Theory to raise issues.

Firstly, Community of Practice: Sterner (2017), from Sweden, focuses on mathematics teachers' participation in a collaborative learning process, specifically on teacher change. Hord, who is an expert on teacher change, is not cited. How do we access what we do not know? Malara and Navarra (2017) use community of practice and co-learning partnerships (Jaworski, 2003) to support primary teachers in developing "adherence to a set of theoretical principles" (p. 2946), which gives an example of a move to a collective of learning that could be analysed using Hord's five principles.

Secondly, Activity Theory: Stouraitis (2017) uses the theory to move between the collective of a "reflection group" of teachers in the same school community and the different teaching actions they undertake in response to issues. Batteau (2017) works with Lesson Study (LS) with combined theoretical perspectives of the French didactical approach and activity theory. This paper moves from the collective LS discussions to analyse one teacher's practices. Tyskerud, Fauskanger, Mosvold and Bjuland (2017) again work with LS, for analysis combining two socio-cultural theories, Activity Theory and Sfard's Commognitive Theory. These examples illustrate the richness of the field and the need for a move to a collective of learning within the field of mathematics education. Hord's work would support the analysis of the developing Professional Learning Communities.

5.3 Theory of learning and/or theory of teaching

Professional Learning Community and Collaborative Learning are among many dozens of contemporary discourses on learning that focus on such similar-sounding themes as *collaboration*, *cooperation*, *collectivity* and *community*. Discourses that foreground such notions share an interest in group-based activity, but they can vary to the point of incompatibility on the core matters of the nature and locus of learning. In some discourses, the group is understood as a *collection of learners*, and so it is treated mainly as a context or device to support individual learning. In others, the group is interpreted as a *collective of learners*, which prompts shifts in focus to the sorts of goals, dynamics, qualities and/or possibilities that can emerge in joint activity. In others, the simultaneity of individual and group transformation is foregrounded, in which the community is recognized as unfolding from and enfolded in the individual.

To our reading, the matter of the simultaneous transformation of individual and group – that is, regarding both the individual and the group as a learner – is a core theme of community-focused theories. While this theme is certainly represented across multiple papers in the CERME10 proceedings, perhaps, somewhat ironically, it seems to come out most powerfully not in specific reports but across the network of reporters. That is, the simultaneity of individual and group learning is especially evident in the authorship of the proceedings, but much less so in the actual accounts provided.

6. Discussion and conclusions

As noted at the start of this chapter, our original charge was to engage with the topic of a theoretical/analytical framework for analysing mathematics teacher education and professional development. We believe it entirely possible to contrive an encompassing framework, despite the broad ranges of contemporary interests and perspectives that would render such a task daunting.

Indeed, based on the issues arising through our cases presented here, we believe that it is time for the field to foreground and support efforts to develop frameworks that afford access to the varied convergences and divergences in the interests and perspectives currently at play in discussions of teacher learning. To our analysis, the coupled influences of a great diversity of theoretical offerings and a broad range of necessary considerations present an ever-growing risk of fragmentation in the field as researchers are compelled to focus more on superficialities than on deep convictions. Such ranges and diversities can enable the development of a domain if they can be brought into productive conversation. However, unexamined variety is more likely to obstruct development in absence of some deliberate means of foregrounding the interests and convictions at play.

To that end, we would propose that a theoretical/analytical framework for analysing mathematics teacher education and professional development should be structured around neither specific theories of learning nor particular research interests, but on the generally more implicit images and metaphors used by researchers and theorists to characterize learning and learners. As is evident in the cases presented in this chapter, attending too closely to explicit foci presents the risk of eclipsing major issues. Two such issues are exemplified in our cases in the often-underappreciated distinctions: *theorizing learning* versus *influencing learning*, and *individual learning/learners* versus *group learning/learners*. We suspect that underattendance to these dyads may be impeding the evolution of the field, as attention channels toward superficial similarities/differences that can mask profound incompatibilities/complementarities.

For example, regarding the pairing of *theorizing learning* and *influencing learning*, there appears to be a marked tendency in the field to conflate theories of learning with theories of teaching. Clearly, it would be ridiculous to suggest that such matters can or should be tidily separated. However, we might argue that it is just as ridiculous to conflate insights into the complex dynamics on human learning with direct advice for teachers. Learning and teaching are complex phenomena. It is rarely possible to map a singular insight about the former into a tidy recommendation for the latter. As with all co-entangled, complex phenomena, affecting one element of a system will reverberate across many elements.

Similarly, while notions of collectivity are currently well represented in the research literature, it is not clear that the dyad of *individual learning/learners* and *group learning/learners* is engaged with in complementary, or even compatible ways across reports. To our analysis, the naming of specific theories, such as Community of Practice or Collaborative Learning, does more to obscure the problem than to clarify it. When literally hundreds of discourses on learning are active in the field, communications will be stifled when titles of theories are permitted to stand in for theoretical convictions – especially when those researchers rely on secondary sources.

Against this backdrop, we are acutely aware that any effort in the direction of a comprehensive theoretical/analytical framework for analysing mathematics teacher education and professional development must respond to criticisms of the distorting influences of modernist efforts to totalize, essentialize or reduce knowledge. An overarching framework, it is commonly argued, will inevitably overemphasize some elements and ignore others. We recognize this legitimate concern. At the same time, however, we are troubled by the results of the analyses presented in this chapter. While an overarching framework will privilege some discourses, there is strong evidence that an absence of a framework is obstructing communications and stunting the development of the field.

What, then, to do? How might a framework be developed that is not only attentive to the sorts of major dyads just discussed but that accommodates interest in such topics as context, motivation, experience, identity, activism and language – to name just a few contemporary research obsessions.

Oriented by the example of the learningdiscourses.com website, we would suggest that such a framework should be more about description than prescription. We suspect that an effort toward description would be broader and more inclusive, affording appreciations of how discourses converge and diverge, what they spotlight, what they obscure and how they cluster around assumptions or interests. In the process, more attention might be given to the often-underappreciated alignments and discontinuities that reside in assumptions and habits of association. To this end, we would argue that efforts toward a theoretical/analytical framework for analysing mathematics teacher education and professional development should not be engaged as attempts to convince or to privilege, but as means to expand understanding and enable productive communication.

References

Adler, J., Ball, D., Krainer, K., Lin, F. L., & Novotna, J. (2005). Reflections on an emerging field: Researching mathematics teacher education. *Educational Studies in Mathematics*, *60*, 359–381.

Anfara, V. A., Jr., & Mertz, N. T. (Eds.). (2015). *Theoretical frameworks in qualitative research* (2nd ed.). Thousand Oaks, CA: Sage Publications.

Baldry, F. (2017, February 1–5). A teacher's orchestration of mathematics in a "typical" classroom. In T. Dooley & G. Gueudet (Eds.), *Proceedings of the Tenth Congress of the European Society for Research in Mathematics Education* (pp. 3041–3048). CERME10. Dublin, Ireland: DCU Institute of Education and ERME.

Batteau, V. (2017, February 1–5). Study of primary school teacher's practices for a lesson after a lesson study process in mathematics. In T. Dooley & G. Gueudet (Eds.), *Proceedings of the Tenth Congress of the European Society for Research in Mathematics Education* (pp. 3256–3263). CERME10. Dublin, Ireland: DCU Institute of Education and ERME.

Björklund, C. (2017, February 1–5). Aspects of numbers challenged in toddlers' play and interaction. In T. Dooley & G. Gueudet (Eds.), *Proceedings of the Tenth Congress of the European Society for Research in Mathematics Education* (pp. 1821–1828). CERME10. Dublin, Ireland: DCU Institute of Education and ERME.

Carrillo, J., Santos, L., Bills, L., & Marchive, A. (2007, February 22–26). From a study of teaching practices to issues in teacher education. In D. Pitta-Pantazi & G. Philippou (Eds.), *Proceedings of the Fifth Congress of the European Society for Research in Mathematics Education* (pp. 1821–1826). CERME5. Larnaca, Cyprus: University of Cyprus, Department of Education and ERME.

Davis, B., & Brown, L. (2009). Development of teaching in and of practice. In R. Even & D. Loewenberg Ball (Eds.), *The professional education and development of teachers of mathematics* (pp. 149–166). 15th ICMI Study. New York, NY: Springer. doi:10.1007/978-0-387-09601-81

Davis, B., & Francis, K. (2019a). Socio-cultural-focused discourses. *Discourses on Learning in Education*. Retrieved from https://learningdiscourses.com

Davis, B., & Francis, K. (2019b). Variation theory. *Discourses on Learning in Education*. Retrieved from https://learningdiscourses.com

Davis, B., & Francis, K. (2019c). Professional learning community. *Discourses on Learning in Education*. Retrieved from https://learningdiscourses.com

Dooley, T., & Aysel, T. (2017, February 1–5). Using variation theory to explore the reteaching phase of lesson study. In T. Dooley & G. Gueudet (Eds.), *Proceedings of the Tenth Congress of the European Society for Research in Mathematics Education* (pp. 3758–3760). CERME10. Dublin, Ireland: DCU Institute of Education and ERME.

Dooley, T., & Gueudet, G. (Eds.). (2017, February 1–5). *Proceedings of the Tenth Congress of the European Society for Research in Mathematics Education*, CERME10. Dublin, Ireland: DCU Institute of Education and ERME. Retrieved December 19, 2019, from https://hal.archives-ouvertes.fr/CERME10

Engeström, Y. (1999). Activity theory and individual and social transformation. In Y. Engeström, R. Miettinen, & R.-L. Punamäki (Eds.), *Perspectives on activity theory*. Cambridge, UK: Cambridge University Press.

Hargreaves, A. (2019). Teacher collaboration: 30 years of research on its nature, forms, limitations and effects. *Teachers and Teaching*, *25*(5), 603–621. doi:10.1080/13540602.2019.1639499

Hord, S. M. (1997). *Professional learning communities: Communities of continuous inquiry and improvement.* Austin, TX: Southwest Educational Development Laboratory. Retrieved from www.sedl.org/pubs/change34/plc-cha34.pdf

Jaworski, B. (2003). Research practice into/influencing mathematics teaching and learning development: Towards a theoretical framework based on co-learning partnerships. *Educational Studies in Mathematics, 54*(2/3), 249–282.

Jaworski, B. (2005). Learning communities in mathematics: Creating an inquiry community between teachers and didacticians. *Research in Mathematics Education, 7*(1), 101–119.

Lave, J., & Wenger, E. (1991). *Situated learning: Legitimate peripheral participation.* Cambridge, UK: Cambridge University Press.

Ling, L., & Marton, F. (2012). Towards a science of the art of teaching: Using variation theory as a guiding principle of pedagogical design. *International Journal for Lesson and Learning Studies, 1*(1), 7–22.

Llinares, S., Krainer, K., & Brown, L. (2018). Mathematics teachers and curricula. In S. Lerman (Editor-in-Chief), & M. Artigue (section editor) (Eds.), *Encyclopedia of mathematics education* (2nd ed.). Springer. https://doi.org/10.1007/978-3-319-77487-9_111-5

Malara, N. A., & Navarra, G. (2017, February 1–5). Integrating teachers institutional and informal mathematics education: The case of "Project ArAl" group in Facebook. In T. Dooley & G. Gueudet (Eds.), *Proceedings of the Tenth Congress of the European Society for Research in Mathematics Education* (pp. 2940–2947). CERME10. Dublin, Ireland: DCU Institute of Education and ERME.

Marton, F. (1970). *Structural dynamics of learning.* Stockholm: Almqvist & Wiksell.

Marton, F. (2015). *Necessary conditions of learning.* New York, NY: Routledge.

Marton, F., & Booth, S. (1997). *Learning and awareness.* Mahwah, NJ: Lawrence Erlbaum Associates.

Marton, F., & Pang, M. F. (2003). Beyond "lesson study": Comparing two ways of facilitating the grasp of economic concepts. *Instructional Science, 31*(3), 175–194.

Marton, F., & Pang, M. F. (2006). On some necessary conditions of learning. *Journal of the Learning Sciences, 15,* 193–220.

Mason, J. (2002). *Researching your own practice: The discipline of noticing.* Abingdon & Oxford: Routledge.

Nardi, E. (2017, February 1–5). From advanced mathematical thinking to university mathematics education: A story of emancipation and enrichment. In T. Dooley & G. Gueudet (Eds.), *Proceedings of the Tenth Congress of the European Society for Research in Mathematics Education* (pp. 9–30). CERME10. Dublin, Ireland: DCU Institute of Education and ERME.

Odindo, F. (2017, February 1–5). Teacher learning in teaching the topic of quadratic expressions and equations in Kenyan high schools: Learning study design. In T. Dooley & G. Gueudet (Eds.), *Proceedings of the Tenth Congress of the European Society for Research in Mathematics Education* (pp. 3129–3136). CERME10. Dublin, Ireland: DCU Institute of Education and ERME.

Robutti, O., Cusi, A., Clark-Wilson, A., Jaworski, B., Chapman, O., Esteley, C., . . . Joubert, M. (2016). ICME international survey on teachers working and learning through collaboration. *ZDM Mathematics Education, 48*(5), 651–690.

Runesson, U. (2005). Beyond discourse and interaction: Variation: A critical aspect for teaching and learning mathematics. *Cambridge Journal of Education, 35*(1), 69–87.

Runnesson Kempe, U., Lövström, A., & Hellqvist, B. (2017, February 1–5). Exploring how "instructional products" from a theory-informed lesson study can be shared and enhance student learning. In T. Dooley & G. Gueudet (Eds.), *Proceedings of the Tenth*

Congress of the European Society for Research in Mathematics Education (pp. 3153–3160). CERME10. Dublin, Ireland: DCU Institute of Education and ERME.

Sakonidis, C., Drageset, O. G., Mosvold, R., Skott, J., & Didem Taylan, R. (2017, February 1–5). Introduction to the papers of TWG19: Mathematics teachers and classroom practices. In T. Dooley & G. Gueudet (Eds.), *Proceedings of the Tenth Congress of the European Society for Research in Mathematics Education* (pp. 3033–3040). CERME10. Dublin, Ireland: DCU Institute of Education and ERME.

Slavit, D. (2020). Frameworks for analysing collaborative teacher activity. In G. M. Lloyd & O. Chapman (Eds.), *International handbook of mathematics teacher education* (Vol. 3, pp. 15–50). Leiden, The Netherlands: Brill Sense.

Sterner, H. (2017, February 1–5). Long-term learning in mathematics teaching and problematizing daily practice. In T. Dooley & G. Gueudet (Eds.), *Proceedings of the Tenth Congress of the European Society for Research in Mathematics Education* (pp. 3169–3176). CERME10. Dublin, Ireland: DCU Institute of Education and ERME.

Stouraitis, K. (2017, February 1–5). Teachers' decisions and the transformation of teaching activity. In T. Dooley & G. Gueudet (Eds.), *Proceedings of the Tenth Congress of the European Society for Research in Mathematics Education* (pp. 3175–3184). CERME10. Dublin, Ireland: DCU Institute of Education and ERME.

Tirosh, D., Tsamir, P., Barkai, R., & Levenson, E. (2017, February 1–5). Preschool teachers' variations when implementing a patterning task. In T. Dooley & G. Gueudet (Eds.), *Proceedings of the Tenth Congress of the European Society for Research in Mathematics Education* (pp. 1917–1924). CERME10. Dublin, Ireland: DCU Institute of Education and ERME.

Tyskerud, A., Fauskanger, J., Mosvold, R., & Bjuland, R. (2017, February 1–5). Investigating lesson study as a practice-based approach to study the development of mathematics teachers' professional practice. In T. Dooley & G. Gueudet (Eds.), *Proceedings of the Tenth Congress of the European Society for Research in Mathematics Education* (pp. 3384–3391). CERME10. Dublin, Ireland: DCU Institute of Education and ERME.

Venkat, H., & Askew, M. (2017, February 1–5). Focusing on the "middle ground" of example spaces in primary mathematics teaching development in South Africa. In T. Dooley & G. Gueudet (Eds.), *Proceedings of the Tenth Congress of the European Society for Research in Mathematics Education* (pp. 3193–3200). CERME10. Dublin, Ireland: DCU Institute of Education and ERME.

Watson, A. (Ed.). (2018). *Variation in mathematics teaching and learning: A collection of writing from ATM mathematics teaching.* Derby, UK: ATM.

Watson, A., & Mason, J. (2005). *Mathematics as a constructive activity: Learners generating examples.* New York, NY: Lawrence Erlbaum Associates.

9

PROMOTING AND INVESTIGATING TEACHERS' PROFESSIONALIZATION PROCESSES TOWARDS NOTICING AND FOSTERING STUDENTS' POTENTIALS

A case of Content-Specific Design Research for Teachers

Susanne Prediger, Susanne Schnell &
Kim-Alexandra Rösike

Noticing and fostering students' mathematical potentials is a major task for every mathematics teacher. However, it is still neglected in professional development (PD), and even underdetermined what teachers need to learn for fulfilling this task. In this chapter, we take the example of this PD content to illustrate what Design Research for teachers can mean as well as how it informs the finer specification of what teachers should learn and how they can learn it. For this, we briefly introduce the main ideas of the research approach Design Research for Teachers in Section 1 and the background of the concrete PD Design Research project in Section 2. Section 3 presents exemplary design and research results which are discussed with respect to general methodological issues of the research approach in Section 4.

1. Adopting Design Research for teachers

1.1 Design Research as an established research methodology with big variety

Design Research is a widely established research methodology for enhancing and investigating students' learning. It is especially strong when two aims are to be combined: (1) designing learning arrangements and (2) investigating the initiated learning processes as well as contributing to local instruction theories (Bakker & van Eerde, 2015). Although Design Research approaches share common characteristics (e.g. interventionist, theory generative, iterative, ecologically valid and practice-oriented, cf. Cobb, Confrey, diSessa, Lehrer & Schauble, 2003), a big variety of approaches exists (cf. the 52 case studies documented in Plomp & Nieveen, 2013). These approaches differ in their reasons for doing Design Research, their types of results, their intended roles of the results for educational innovation, their scales and their background theories (cf. Prediger, Gravemeijer & Confrey, 2015a).

Our Dortmund research group follows a topic-specific approach which allows to account for different mathematical topics in detail (Prediger & Zwetzschler, 2013) with a focus on learning processes (ibid.; Prediger et al., 2015a). In the last years, this approach was adapted to designing and researching environments for teacher learning in professional development (Prediger, Schnell & Rösike, 2016; Prediger, 2019).

1.2 Adopting Design Research for many teachers, not only with some

Zawojewski, Chamberlin, Hjalmarson and Lewis (2008) suggested extending the research methodology of Design Research from students to teachers' professional development (PD) "in order to understand both, how teachers develop in their practice and how to design environments and situations to encourage the development of that practice" (Zawojewski et al., 2008, p. 220). Meanwhile, many teacher educators have described impressive individual professionalization effects of *Design Research with teachers* for the exclusive minority of teachers who are privileged to be part of Design Research teams (Smit & van Eerde, 2011; Bannan-Ritland, 2008). Although this is undoubtedly a very intensive PD setting, it is not realizable for scaling up, since many teachers have no access to such an intensive collaboration with researchers. However, scaling up for reaching many teachers throughout the whole of Germany is the critical long-term goal for the authors' work in the DZLM, the German National Center for Mathematics Teacher Education (Rösken-Winter, Hoyles & Blömeke, 2015).

Thus, Prediger et al. (2016) have suggested complementing the approach of Design Research *with some* teachers by *Design Research for many teachers*, taking into account that professional development for scaling up requires well-founded, robust designs for classrooms and PD courses (Burkhardt, 2006; Swan, 2007). Whereas the individual PD work of *researchers* with a selected group of teachers can be based on spontaneous, intuitive decisions in deep discussions, a robust design for PD conducted by *other facilitators* also needs to be grounded on a solid theoretical base, which allows to anticipate possible challenges of the content to be learnt and teachers' typical professionalization processes. This calls for the next two characteristics, content-specificity and process-focus.

1.3 Content-specificity and focus on teachers' professionalization processes

So far, the growing body of research on conditions and effects of PD is mainly focused on *pedagogical principles* for PD programmes (e.g. Timperley, Wilson, Barrar & Fung, 2007). But for robust designs for scaling up, a good theoretical base for the *content* of the PD course itself also is relevant, which cannot be taken for granted (Prediger, Quasthoff, Vogler & Heller, 2015b). Specifying what teachers should learn about a certain content (e.g. a mathematical topic or noticing students' difficulties) is of course an a priori task before starting the PD. It usually refers to the current state of research on classroom practices or teachers' professional knowledge for this content. However, the further refinement of the content is also an empirical task, as the original content must be restructured with respect to typical teachers' perspectives, which can be reconstructed when qualitatively investigating content-specific professionalization processes.

In their review on PD research, Goldsmith, Doerr and Lewis (2014) emphasize the need to focus on teachers' professionalization *processes* rather than only on quantitatively measurable effects. Even if they have not found much research on processes yet, they collect indications that teacher learning "is often incremental, nonlinear, and iterative, proceeding through repeated cycles of inquiry" (ibid, p. 20). As the research gap is even bigger for *content-specific* research results, a major aim of the approach presented here is to provide fine-grained insights into teachers' processes of professionalization on different specific PD contents. For this aim, the most appropriate approach is the adaptation of topic-specific Design Research with a focus on learning processes (elaborated for classrooms in Prediger & Zwetzschler, 2013; Prediger et al., 2015a). Adapted to the level of teachers, we call it *Design Research for teachers with a focus on content-specific professionalization processes*.

1.4 Four intertwined working areas for PD Design Research

Figure 9.1 shows the four working areas that are *iteratively* connected in the design and research process, adapted from Prediger and Zwetzschler (2013) for PD Design Research by Prediger et al. (2016). The four working areas comprise: (1) specifying and structuring PD goals and contents in hypothetical intended professionalization trajectories; (2) developing the specific PD design; (3) conducting and analyzing design experiments in PD settings and (4) developing contributions to local theories on professionalization processes.

The areas are *intertwined* in the sense that each cycle builds upon results of previous cycles across the areas. Corresponding to the two general aims of Design Research, design results and research results have equal importance: The design results comprise the PD course settings as well as their backgrounds, a specified and structured PD content and refined design principles. The local theories are

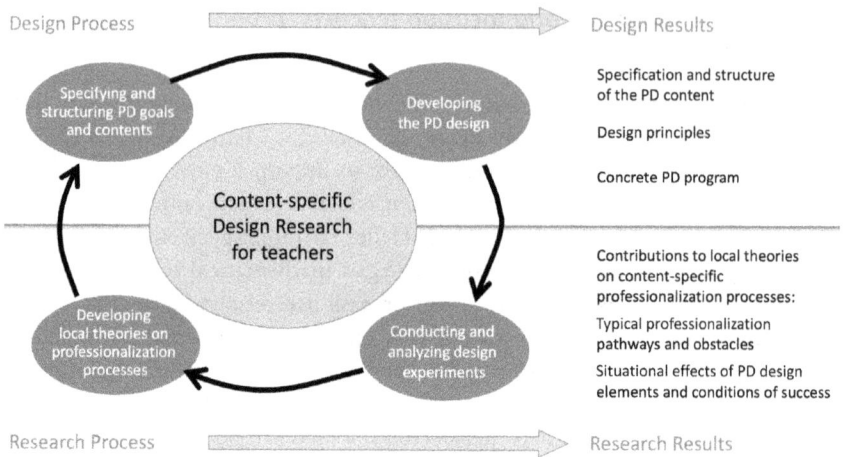

FIGURE 9.1 Working areas and results of Content-Specific Design Research for Teachers.

developed to underpin the concrete products and to be generalized by accumulation over several projects. Contributions to local theories on content-specific professionalization processes can be expected with respect to typical individual pathways and obstacles, means for support in the PD setting as well as their effects and contextual conditions of success.

2. The case of DoMath, a PD Design Research project on noticing students' potentials

For illustrating the approach, we briefly give some insights into the dual Design Research project DoMath (working on student and teacher levels, here focused on the teacher level). The project addresses secondary school mathematics teachers who intend to develop their competences of noticing and fostering students' mathematical potential. Due to space limitations, we focus mainly on noticing rather than fostering.

2.1 Goals, structure and background of the DoMath PD programme

Goals of the classroom programme

The DoMath classroom programme aims at fostering students' mathematical potential. Specifically, it addresses (often underprivileged) students "who can achieve a high level of mathematical performance when their potential is realized to the greatest extent" (Leikin, 2009, p. 388), but have not yet been identified as talented (Suh & Fulginiti, 2011). Thus, the programme adopts a broader conceptualizing of mathematical potential as (1) *potentially hidden*, appearing not only in constantly good performances but also situationally and in "seeds', which have to be fostered in order to be fully realized; (2) *dynamic*, so that it can be fostered over different situations rather than as a stable, naturally given talent, and (3) *participatory*, referring to approximately 20% of all students rather than the top 1% or 2% (Schnell & Prediger, 2017). According to a literature review, mathematical potential can be identified by five different facets (cf. Table 9.1; Leikin, 2011; Sheffield, 2003).

The instructional design therefore builds upon whole-class enrichment settings with rich, self-differentiating open-ended problems (Sheffield, 2003). One example for such an open-ended problem is the so-called Stair Number Exploration (cf. Figure 9.2).

These kind of open-ended problems intend to allow teachers to notice early seeds of students' potentials in the rich situations and to adaptively foster the noticed situative potentials by supportive interactions.

TABLE 9.1 Excerpt of facets of mathematical potential.

Cognitive facet	Meta-cognitive facet	Personal & affective facet	Communicative & linguistic facet	Social facet
• Conceptual understanding • Procedural fluency • Adaptive (logical) reasoning • Finding pattern • Generalizing • Etc.	• Planning • Monitoring • Evaluating	• Mathematical self-concept • Commitment • Creativity • Etc.	• Complex argumentation • Deep discursive involvement • Etc.	• Cooperative skills • Social involvement • Etc.

Stair Number Exploration

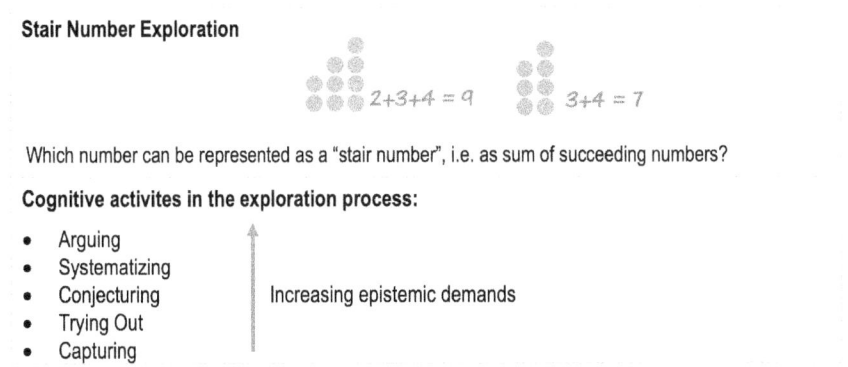

2+3+4 = 9   3+4 = 7

Which number can be represented as a "stair number", i.e. as sum of succeeding numbers?

Cognitive activites in the exploration process:

- Arguing
- Systematizing
- Conjecturing
- Trying Out
- Capturing

Increasing epistemic demands

FIGURE 9.2 Stair number exploration as example for a self-differentiating, open-ended problem.

Structure of the PD programme

Based on the DoMath classroom programme, the DoMath PD programme spans over several months and professionalizes teachers in action and reflection settings of material-based video clubs (Sherin & van Es, 2009). In the PD sessions, teachers reflect on classroom video clips and student products stemming from their teaching experiments with the jointly prepared whole-class enrichment settings (Rösike & Schnell, 2017). The preparation includes their own mathematical inquiries as well as anticipating students' ideas. The typical "sandwich structure" of the PD programme is illustrated in Figure 9.3.

Background

The general PD content *noticing* has been characterized in several research studies: They emphasize the need for teachers to overcome deficit-oriented modes on students and the necessity of a shift from product- to process-oriented perspectives (Empson & Jacobs, 2008). By the construct of professional vision, Sherin and van Es (2009) conceptualize noticing by three subconstructs: (1) selective attention, (2) knowledge-based reasoning and (3) interpreting specific events in terms of broader pedagogical principles.

In the specific case of noticing students' mathematical potentials with a dynamic and participatory conceptualization of potential, all three subconstructs are important: For uncovering hidden potentials, the process perspective in a non-deficit-oriented mode is hypothesized to be an important precursor for extending the selective attention and widening the repertoire of possible actions for fostering the fragile situative potentials (cf. Figure 9.4 for the intended

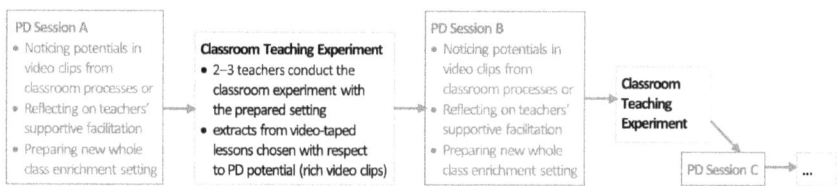

FIGURE 9.3   Structure of the PD programme with sessions and intermediate classroom experiments.

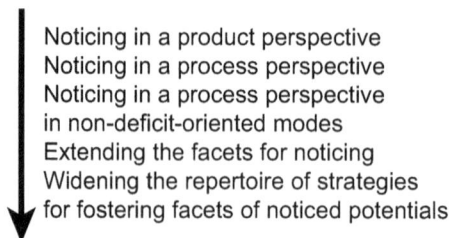

Noticing in a product perspective
Noticing in a process perspective
Noticing in a process perspective
in non-deficit-oriented modes
Extending the facets for noticing
Widening the repertoire of strategies
for fostering facets of noticed potentials

FIGURE 9.4   Intended professionalization trajectory.

professionalization trajectory which corresponds to a hypothetical learning trajectory in other Design Research approaches; cf. Prediger et al., 2015a).

2.2 Project design in three iterative cycles with mini cycles

Overall project design

The DoMath PD programme is developed and investigated in an ongoing PD Design Research project in the described approach (cf. Section 1) from 2014 to 2018. Three iterative cycles of design experiment series were conducted in 2014–2015 (with 5 teachers in 6 PD sessions over 12 months), 2015–2016 (with 20 teachers in 6 PD sessions over 10 months) and 2016–2017 (with 23 teachers in 3 whole-day PD sessions of 2 days each, over 6 months). During each design experiment cycle, the PD sessions and classroom teaching experiments (cf. Figure 9.3) were investigated in so-called mini-cycles of Design Research for immediately refining the programme. Analyses of teachers' discussions and their subsequent implementation of activities in the classroom informed the design of the following PD sessions, so that overall the relatively vague intended trajectory matured into a more detailed model for noticing students' potentials (Schnell & Prediger, 2017). Later, this refinement of the underlying content-specific theory will allow pursuing the long-term aim to develop a PD course for scaling up to a national level with facilitators within the DZLM.

Methods for data gathering

The data collection includes video recordings of classroom teaching experiments and all PD sessions (cf. Figure 9.3). In order to capture teachers' perspectives, data from group discussions were complemented by interviews with single teachers. These individual interview sessions in which video clips are analysed allow deeper insights into the individual professional visions of the teachers.

Methods for data analysis

The qualitative methods for analyzing transcripts from the video data aimed at (1) specifying demands and challenges in teachers' noticing and (2) reconstructing individual pathways in professionalizing the noticing. For this, video data of PD sequences in which teachers discussed the potential of students were selected and analused by means of the sensitizing subconstructs of professional vision (I–III) and the theoretically derived facets for identifying potentials (Table 9.1) in an inductive-deductive procedure. Comparing the sequences of the video clip selected by the teachers (I: selective attention) with a pre-analysis of the clips conducted by the research team (cf. Rösike & Schnell, 2017) was used to identify the challenges in noticing potential. Next, the selected sections were interpreted and carefully discussed in terms of the underlying perspective on potential (inductively specifying the PD content; cf. Figures 9.4 and 9.5). The excerpts presented here stem from the ongoing analysis of professionalization contents and pathways of Design Experiment Cycle 2 and are based on 33 hours of video material (13 hours PD sessions, 16 hours of their classroom interactions, 4 hours individual discussions of videos in Cycle 2).

3. Exemplary design and research results in the DoMath project

In the following, we present exemplary insight into some results of the Design Research project focusing on teachers' professional development. Regarding the *design*, we briefly discuss two of the design elements implemented, evaluated and successively adapted over the course of the design cycles. In terms of the *research* results, we discuss obstacles in the teachers' professionalization process which lead to refinements of the local theory for noticing and fostering students' mathematical potential. As both types of results are intertwined, they are described first before being illustrated by an exemplary analysis of a transcript.

3.1 Design Results

Design element video clips from participants' classrooms

In line with other professionalization approaches (e.g. Sherin & van Es, 2009; Empson & Jacobs, 2008), the project uses video clips from the participating teachers' classrooms which are intensively discussed by groups of teachers. The collective video analysis aims at promoting discussions among the teachers, in which they deepen their own understanding of mathematical potential by collectively negotiating which student utterances or actions can be interpreted as situational seeds of potential and which interactions by the teachers could have possibly lead to fostering them (for a detailed description of a video clip cf. Rösike & Schnell, 2017). Even though the analysis takes place in a laboratory setting to allow for in-depth inquiries, the clips are taken from real classroom activities to facilitate the transferability of acquired noticing skills into the teachers' everyday practice. The clips are selected by the research team in terms of student interactions showing high levels of cognitive activities, productive discussions or other seed of mathematical potential. However, the captured working processes are often non-linear and can be interspersed with less productive phases or incorrect approaches, which allow for a large variety of analytic perspectives.

Design elements focus question and diagnostic categories

The collective discussion of classroom video clips allows for a variety of settings such as letting the teachers pick relevant scenes themselves vs. pre-picked scenes or developing an analysis focus during the meeting vs. pre-given focus questions (cf. van Es & Sherin, 2006). Results of van Es and Sherin (2006) point out that a less focused discussion enables the teachers to adopt more diverse perspectives with a broader scope of different aspects that are considered. A pre-given focus question, on the other hand, leads to a narrower perspective with insights on a more in-depth level, e.g. into student thinking. Both – the narrow as well as the broad perspective – are valuable for a comprehensive account of the classroom interactions.

In the project DoMath, the first video club meetings in Design Experiment Cycle 1 did not immediately aim at directing teachers' noticing of potential. Instead, the first goal was to identify in-service teachers' conceptions of mathematical potential as a starting point to complement the theoretically derived facets and aspects (cf. Table 9.1). However, teachers kept a deficit-oriented mode for a long time, which hindered them in seeing any seeds of mathematical potential at all. Thus, after the first mini cycles, we introduced the focus question "What kind of potentials can you discover in the students' processes?" which was repeated by the researchers throughout the discussion session. Furthermore, to facilitate the focus on potential under the cognitive facet (cf. Table 9.1), additional scaffolding material was provided which included didactical categories for grasping typical cognitive processes when working on open-ended tasks. For example, the Stair Number Exploration in Figure 9.2 was discussed together with the categories for cognitive activities in that figure.

As the following analysis will exemplify, the effect of the focus question and categories was substantial in the second cycle: from the first PD session on, the second teacher group adopted more of a process perspective in mainly non-deficit modes. The categories allowed the teachers to focus their attention on the processes rather than the products and give them a language to distinguish different moments in a refined way.

3.2 Exemplary research results: extending the landscape of diagnostic perspectives and their relation to fostering

To illustrate perspectives and obstacles during a typical collective discussion of the video clips, we present short excerpts from the PD Session 2 in Cycle 2. Afterwards, we condense the findings into the developed perspective model for noticing and fostering potentials.

Functioning of focus questions and categories

In the following transcript, the teachers watch a video clip – complemented by the transcript – of four students working on the Stair Number Exploration in Figure 9.2 which was recorded in a Grade 8 classroom taught by the participating teacher Katharina. Next, they are asked to comment the scene spontaneously and make sense of the students' (non-linear) exploration process in groups of two or three. Lastly, the facilitator and researcher (the second author) asks them to focus on mathematical potential, which is then discussed in the whole group. The transcript was translated from German and simplified marginally to enhance the readability. All names are pseudonyms.

41 FACILITATOR: Let's bring it all together. What did you notice? Where are the potentials, which cognitive activities do you see? . . .

42 JULIA: I go first. The first part, the first 43 seconds. It seems like they try something systematically. Though, in our opinion, Kai is quite dominant. And Lukas is still stuck at understanding [the task]. In between he yells "no, again, again, again", he is not there, yet. And in the end "ah yes, you can do it with 7, that's correct". So he is a bit behind the others and as I said, Kai dominates it all. Until there is the first conjecture after 43 seconds. . . .

46 SVENJA: In our group, we wondered if the students were on the same page regarding if they look at stairs with two steps or stairs with three steps. When you only read the transcript, we got the feeling that one thinks about stairs with three steps in the beginning and another about stairs with two steps and they are not on the same page. Later, they agree with each other: when the first conjecture is verbalized, then they find each other and notice that they have to get on the same page. . . .

While the facilitator's focus question in Line 41 addresses the mathematical potential in the scene, the teachers Julia and Svenja mainly concentrate on understanding and evaluating how the students cope with the problem. Both point out deficits in the group working process as one student is lost and the others are perceived to talk about different things in the beginning. This perspective of evaluating if the task is addressed correctly by the students is quite prominent in everyday classrooms, as the teacher has to make sure that everyone is able to understand and work on the given assignment. Svenja's last sentence can be understood as a more positive evaluation as she points out how the students overcome their difficulties themselves, thus taking a more non–deficit–oriented stance.

48 SOPHIE: I find that they begin quite quickly to systematize and justify. When they find out that all uneven numbers work, they all confirm each other one after the other. They say "then you always have stairs" in line 23 [*of the transcript*] for example . . . and again in line 33. Confirms "because you only have one step", which is for them already a justification that all uneven numbers work. And then they continue immediately: next you have to mathematize the even numbers. Here, they are also very fast to try and find a generalization.

55 SVENJA: [*Referring to the given categories of typical activities in exploration processes*] I think they always jump a step up and then another and then one back and then again one up. They always get a step further and then one step back until they justify it at some point. Not on the first pages [of the transcript], I wouldn't say they justify mathematically correct yet, written or in any other way. At least they did go many steps up, as you said (*to Merle*), but also down. It's a constant movement. Nothing written down and never staying in the same spot, but always moving.

These statements by Sophie and Svenja show that they now apply the previously introduced categories for exploration processes (in Figure 9.3) as a lens to interpret the students' working process, which leads them to an inquiry of their cognitive activities. Thus, the design element *focus question and categories* seems to successfully initiate the analysis of a more in-depth level of student thinking. Furthermore, Svenja comments on the complexity and non-linearity of the process which surfaces only by the design element *video clip*. The diversity of aspects and perspectives mentioned by the teachers shows the potential of the selected video clips for activating teachers' diverse diagnostic processes.

While these statements above seem to be non-deficit oriented, it remains unclear if the teachers already perceive the situational mathematical potential. Thus, the researcher reminds the group of the focus question.

72 FACILITATOR: Did you notice anything else where you would say "There is something I would call potential. Something is happening which is mathematical potential for me"?

73 EMMA: Me and my colleague, we first had the impression that Tom just listens in the beginning. And I tried to find out if he works on the task or if he thinks about something else. And then we read the transcript and my colleague found out that we have to look more closely, because Tom is always structuring a bit. He lets others think and calculate and then he gives a small summary of what the others did. I find that pretty interesting. I wouldn't have thought so while watching the scene, but by looking so closely. . . . It is really the case, Tom is not absent or lazy and letting the others calculate, but he seems to do something completely different. He thinks about what is the common thing which they all just discovered and then he summarizes it in a structured way. I find that super interesting to discover.

74 CHRISTIAN: He refuses a bit to try out and always stays on a higher level and looks down what the others do. And then he looks up to elaborate on that.

77 MERLE: And in line 85, when they are at the beginning of finding a term. It was about having to subtract six, which was true for the stairs with three steps. And then he says "or minus 10" which is a reproduction, because it was said before, but I think it's really clever to say it in this moment. Because he says you always have to subtract the base form, and the base form is always different. It can be 6, it can be 10. And thus they figure out that they have to subtract [a variable] z.

(Cycle 2, PD session 2, Clip "Stair Number Exploration",
min. 82:49–99:37)

In line 73, Emma describes how her perception of the student Tom changed: While she was first unsure if he was working on the task at all (i.e. identifying a possible deficit in not being engaged in the working process), the detailed analysis of the transcripts leads her to a quite positive evaluation of Tom as a moderator who repeatedly summarizes, systematizes and thus pushes the exploration process of the group forward. Christian and Merle also point out the high cognitive level of Tom's contributions. A possible interpretation of these statements is that the teachers were able to achieve a more in-depth level of analysis which leads them to uncover a potential in this situation, which they ascribe to Tom. Again, the application of hierarchical categories support them in their positive evaluation. While this potential might have been hidden in the very beginning of the clip, it becomes apparent in the student's actions later on.

Accounting for obstacles and teachers' perspectives

The described snapshots from the second PD meeting in the second cycle are prototypical for teachers' differences and obstacles in changing perspectives: Typically, teachers at first argue from a deficit-oriented mode which is overcome by the focus questions. However, the process perspective does not automatically lead to focusing hidden potentials and searching for strategies to foster them. Instead of thinking about strategies to foster uncovered seeds of situational potential, the teachers identify and discuss mainly strategies to help students to solve the open-ended mathematical problems. In consequence, the noticing focuses primarily on students' processes of coping with the task (or why they could not cope well). This can be illustrated in an even more pointed way by the following excerpts of data.

After watching a video clip of two female students working on an open task about several derivatives (Grade 12), Sonja, one of the video-watching teachers in the third PD session, says

78 SONJA: Where they have problems is with verbalizing what they found out – especially mathematically correct verbalizing. So, I think they did understand the principle, but [not the relevant pattern behind it].
And well, you have to justify or formulate it in a more differentiated way.
(Cycle 2, PD Session 3, Clip "Derivatives", min. 16:48)

Within her analysis of the video clip, Sonja points out what the girls would have needed to accomplish the problem. She emphasizes what they reached and the discursive obstacles they need to overcome. Sonja's perspective is an instance of what we researchers later decided to call the *process-coping perspective* (see later in this section): although already overcoming purely deficit-oriented modes and focusing on processes, Sonja does not yet focus on potentials. As our teachers often adopt this perspective, we needed to include it into the model and consider it as rational choice, since it is teachers' responsibility to support the students in coping with the task (or their acquisition of competences or knowledge). Hence, it is also a direct successor of the product perspective.

This process-coping perspective often coexists with the *potential indicator perspective* which we have reconstructed when the teacher implicitly poses to her- or himself questions like "Which situational indicators for students' potentials can we identify?" For example, the teacher Stephanie analysed a video clip of four students (Grade 8) working on a problem-solving task:

45 STEPHANIE: That is really a good way of abstraction. They generalize very well at this point. Also, how they stay at it. They know now, they have the odd numbers and now they think about how to adjust the stairs [of numbers]. . . . Thus, they communicate well with one another and then generalize really well. There is a lot of potential.

(Cycle 2, Individual discussion of video clip "Stair Number Exploration", min. 18:42)

Stephanie also reconstructs steps in the coping perspective, but beyond that, she identifies the students' way of abstraction as an indicator for their mathematical potential. At the same time, the way she and some colleagues talk about the students indicates that she conceives potential here as students' stable disposition rather than dynamically emerging and disappearing in the situation which requires teacher's efforts to be stabilized.

It was a longer discussion in the research team to reconstruct the backgrounds for these observed obstacles. After having also re-analysed other transcripts, we realized the need to differentiate the process perspective which is still too vague in the hypothesized learning trajectory (Figure 9.4). The result of several reconstructions and discussions was a refined perspective model (Figure 9.5) which allows to take into account the teachers' perspective and to structuring of the PD content which was not adequately grasped by the earlier learning trajectory in Figure 9.4 (cf. Schnell & Prediger, 2017).

The last perspective, at which the PD programmes aim, is now called the *potential-enhancing perspective*, looking for fragile situational potentials which are worth being strengthened in order to stabilize them. This perspective allows fostering potentials, but in the beginning, teachers rarely adopt this perspective. Nicole is one of the teachers who adapt *potential-enhancing perspective* in later stages of the PD, as her following utterance shows:

334 NICOLE: I would strengthen their *generalizing*, because I think, Aishe is practically formulating a kind of rule. "I thought that in this case, you always do such and such . . ." She is *on the way to systematize*, to find a rule. And I think, well, . . . she has potential in that idea and *we could work on that*.

(Cycle 3, PD Session 1, Day 2, group discussion, min. 2:30:00)

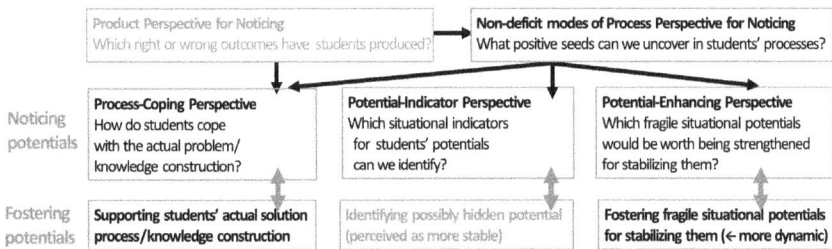

FIGURE 9.5 Refined structure of PD content: perspective model for noticing and fostering potentials.

Also, Henry starts to adopt the *potential-enhancing perspective* and even explains what he should NOT do in order to foster the situational potential:

79 FACILITATOR: Would you have liked to give them an impulse, if you would have been there?

80 HENRY: Yes, I do find it great. So I noticed for myself that it works quite well even if I don't give any prompt. I notice that I, as teacher, would have quickly felt the need to say "oh, look here, what happens here? The three here". And now I think you sometimes give them too little time so that they can unfold their ideas in peace. That it needs a lot of time. . . . Because I find they gave the right impulses themselves.

*(Cycle 2, Individual discussion 2 of video clip
"Stair problem", min. 14:55)*

In total, the research contributed to refining the model as it revealed specifically the following observation on the tight connection between teachers' noticing and fostering:

- What teachers selectively notice is highly connected to what they intend to foster. As long as the main goal is supporting students' actual processes of working on a given task, it is rational to stay in a process–coping perspective (cf. Figure 9.5).
- The potential-indicator perspective looks at indicators for students' existing potentials displayed in a certain situation. While it is important in our teaching approach, it cannot help fostering students when potentials are perceived as pre-existing and more stable dispositions.
- In contrast, sensitive strategies for fostering (still fragile) situational potentials in order to stabilize them in the long run require a potential-enhancing perspective of noticing.

It is this perspective which teachers adopt the least often in the beginning of the course and successively learn to adopt during the discussion of fostering strategies. Rather than linear, teachers' navigation during the professionalization process is forward and backward, since they need to coordinate different perspectives at the same time.

3.3 Summarizing and combining the design and research results

By the case of the DoMath project, we can exemplify typical design and research results of typical PD Design Research projects as listed in Figure 9.1.

Research results

Although the existing literature provided consolidated knowledge of the general structure of teachers' noticing and general pedagogical principles for enhancing them (Sherin & van Es, 2009; Blomberg, Renkl, Sherin, Borko & Seidel, 2013), little was known about the specific content, noticing students' hidden mathematical potentials based on our dynamic and participatory conceptualization of potential and their connection to fostering practices. Thus, the empirical research on teachers' processes was necessary to iteratively refine a local theory on this PD content and individual pathways to approaching it. First research results are condensed in the *perspective model for noticing and fostering potentials* (cf. Figure 9.5). It provides a content-dependent language for describing typical professionalization pathways and obstacles. Of course, the reconstructed insights into effects of specific design elements like focus questions and categories are not yet generalizable. Hence, their transferability to other contents should be investigated in further research.

Design results

The research results on effects of specific design elements have iteratively influenced the design of the DoMath PD sessions within the mini cycles and between the big cycles. However, we have only achieved first steps for the long-term goal of designing a *PD programme with robust materials* that can be used for scaling up, i.e. for facilitators who have not joined our programmes themselves. For this purpose, the theoretical foundation is crucial, and in this sense, the *specification and structure of the PD content* based on the perspective model is also an important design result which will guide a manual for facilitators. With respect to pedagogical *design principles*, the project has mainly confirmed existing work (e.g. Blomberg et al., 2013) and found content-specific ways for their realization, a design result which is far from trivial.

4. Zooming out: discussing the research approach Design Research

Although Design Research *with* teachers on the student level is an excellent setting for professionalizing *some* teachers, this chapter pleads for extending the approach for reaching *many* teachers. In the presented approach, design experiments take place in PD sessions, not in classrooms alone. PD Design Research adds to usual PD programme development a much more intense, video-based analysis of teachers' professionalization pathways during and between the PD sessions, by own teaching experiments and their video-based reflection in small groups. The reconstruction of teachers' individual professionalization pathways allows gaining profound insights into the structure of the PD content: in our case, the process perspective had to be split for understanding teachers' pathways (cf. the perspective model in Figure 9.5).

As with every investigation of individual learning pathways, such an analysis always has the risk of being deficit-focused as the intended goal is not reached, devaluing the perspective of the learning teachers. In order to avoid this, the teachers' perspectives as well as their inner logic and rationalization has to be systematically taken into account. The research goal has to be the search for a synthesis between teachers' and intended perspectives which leads to overcoming the risk of deficit orientation (Prediger et al., 2015b). In our case, we had to accept the process-coping perspective as a natural and important perspective for in-service teachers which should coexist with the potential-enhancing perspective.

The methodological control of the interpretative data analysis procedures is paramount for achieving profound empirical results. This means respecting the quality criteria of transparency, intersubjectivity and openness for phenomena outside the theoretical input. However, quality criteria in Design Research go beyond these classical methodological criteria, as they also comprise relevance and practicability of the design, generalizability of the results by accumulating over several projects and ecological validity of the complete setting (Cobb et al., 2003). For the concrete project, the generalizability of the research results is not yet achieved since the process is only at the beginning. However, its preliminary results are encouraging to pursue this aim.

Funding information

DoMath was originally funded by the Dortmund Stiftung (grant to S. Prediger). We thank the foundation as well as the participating schools and classes and our colleagues and research students for the collaboration. The research is conducted within the DZLM, the German Center for Mathematics Teacher Education (funded by the German Telekom Foundation, grant to S. Prediger).

References

Bakker, A., & van Eerde, D. (2015). An introduction to design-based research. In A. Bikner-Ahsbahs, C. Knipping, & N. Presmeg (Eds.), *Approaches to qualitative research in mathematics education* (pp. 429–466). Dordrecht: Springer.

Bannan-Ritland, B. (2008). Teacher design research: An emerging paradigm. In A. Kelly, R. Lesh, & J. Baek (Eds.), *Handbook of design research methods in education* (pp. 246–262). New York, NY: Routledge.

Blomberg, G., Renkl, A., Sherin., M. G., Borko, H., & Seidel, T. (2013). Five research-based heuristics for using video in pre-service teacher education. *Journal for Educational Research Online, 5*(1), 90–114.

Burkhardt, H. (2006). From design research to large-scale impact: Engineering research in education. In J. van den Akker, K. Gravemeijer, S. McKenney, & N. Nieveen (Eds.), *Educational design research* (pp. 121–150). London: Routledge.

Cobb, P., Confrey, J., diSessa, A., Lehrer, R., & Schauble, L. (2003). Design experiments in educational research. *Educational Researcher, 32*(1), 9–13.

Empson, S. B., & Jacobs, V. J. (2008). Learning to listen to children's mathematics. In T. Wood & P. Sullivan (Eds.), *International handbook of mathematics teacher education* (Vol. 1, pp. 257–281). Rotterdam, The Netherlands: Sense Publishers.

Goldsmith, L., Doerr, H., & Lewis, C. (2014). Mathematics teachers' learning: A conceptual framework and synthesis of research. *Journal of Mathematics Teacher Education*, *17*(1), 5–36.

Leikin, R. (2009). Bridging research and theory in mathematics education with research and theory in creativity and giftedness. In R. Leikin, A. Berman, & B. Koichu (Eds.), *Creativity in mathematics and the education of gifted students* (pp. 383–409). Rotterdam, The Netherlands: Sense Publishers.

Leikin, R. (2011). The education of mathematically gifted students: Some complexities and questions. *The Montana Mathematics Enthusiast*, *8*(1&2), 167–188.

Plomp, T., & Nieveen, N. (Eds.). (2013). *Educational design research: Illustrative cases*. Enschede: SLO.

Prediger, S. (2019). Investigating and promoting teachers' pathways towards expertise for language-responsive mathematics teaching. *Mathematics Education Research Journal*, *31*(4), 367–392. doi:10.1007/s13394-019-00258-1

Prediger, S., Gravemeijer, K., & Confrey, J. (2015a). Design research with a focus on learning processes. *ZDM Mathematics Education*, *47*(6), 877–891.

Prediger, S., Schnell, S., & Rösike, K.-A. (2016). Design research with a focus on content-specific professionalization processes: The case of noticing students' potentials. In S. Zehetmeier, B. Rösken-Winter, D. Potari, & M. Ribeiro (Eds.), *Proceedings of the Third ERME Topic Conference on Mathematics Teaching, Resources and Teacher Professional Development* (pp. 96–105). Berlin: Humboldt-Universität & HAL Archive.

Prediger, S., Quasthoff, U., Vogler, A.-M., & Heller, V. (2015b). How to elaborate what teachers should learn? *Journal für Mathematik-Didaktik*, *36*(2), 233–257.

Prediger, S., & Zwetzschler, L. (2013). Topic-specific design research with a focus on learning processes. In T. Plomp & N. Nieveen (Eds.), *Educational design research: Illustrative cases* (pp. 407–424). Enschede: SLO.

Rösike, K.-A., & Schnell, S. (2017). Do math! Lehrkräfte professionalisieren für das Erkennen und Fördern von Potenzialen. In J. Leuders, T. Leuders, S. Prediger, & S. Ruwisch (Eds.), *Mit Heterogenität im Mathematikunterricht umgehen lernen* (pp. 223–233). Wiesbaden: Springer.

Rösken-Winter, B., Hoyles, C., & Blömeke, S. (2015). Evidence-based CPD: Scaling up sustainable interventions. *ZDM Mathematics Education*, *47*(1), 1–12.

Schnell, S., & Prediger, S. (2017). Fostering and noticing mathematical potentials of underprivileged students as an issue of equity. *Eurasia Journal of Mathematics, Science and Technology Education*, *13*(1), 143–165.

Sheffield, L. J. (2003). *Extending the challenge in mathematics: Developing mathematical promise in K-8 students*. Thousand Oaks: Corwin Press.

Sherin, M. G., & van Es, E. A. (2009). Effects of video club participation on teachers' professional vision. *Journal of Teacher Education*, *60*(1), 20–37.

Smit, J., & van Eerde, H. A. A. (2011). A teacher's learning process in dual design research: Learning to scaffold. *ZDM Mathematics Education*, *43*(6–7), 889–900.

Suh, J., & Fulginiti, K. (2011). Developing mathematical potential in underrepresented populations through problem solving, mathematical discourse and algebraic reasoning. In B. Sriraman & K. Lee (Eds.), *The elements of creativity and giftedness in mathematics* (Vol. 1, pp. 67–79). Rotterdam, The Netherlands: Sense Publishers.

Swan, M. (2007). The impact of task-based professional development on teachers' practices and beliefs: A design research study. *Journal of Mathematics Teacher Education*, *10*(4–6), 217–237.

Timperley, H., Wilson, A., Barrar, H., & Fung, I. (2007). *Teacher professional learning: Best evidence synthesis iteration.* Wellington: Ministry of Education.

van Es, E. A., & Sherin, M. G. (2006). How different video club designs support teachers in "learning to notice". *Journal of Computing in Teacher Education, 22*(4), 125–135.

Zawojewski, J. S., Chamberlin, M., Hjalmarson, M. A., & Lewis, C. (2008). Designing design studies for professional development in mathematics education. In A. E. Kelly, R. Lesh, & J. Baek (Eds.), *Handbook of design research methods in education* (pp. 219–245). New York, NY: Routledge.

10

THEORY-BASED DESIGN OF PROFESSIONAL DEVELOPMENT FOR UPPER SECONDARY TEACHERS – FOCUSING ON THE CONTENT-SPECIFIC USE OF DIGITAL TOOLS

Bärbel Barzel & Rolf Biehler

1. Introduction and overview

The chapter illustrates the research and development approach of the DZLM by focusing on two PD courses for upper secondary teachers, for which the authors of this chapter were responsible. The DZLM has developed standards for

- The general design of the teaching and learning processes in PD courses (design principles), based on research on successful PD courses.
- The design of the content of a PD course, based on a general competence and knowledge facet model for teachers and on research on the specific content of the course such as the use of digital tools or the teaching and learning of probability.

Related to that, a research agenda has been set up:

- Accompanying research and evaluation, which addresses the chain of impact from assessing teacher beliefs and competences, over the self-efficacy regarding teaching a topic, to assess their classroom teaching and students' learning outcome.
- General principles have to be interpreted and concretized in every particular PD course and its boundary conditions.

For all programmes of professionalization – so also for DZLM – the main challenge is the issue of scaling (Coburn, 2003). Therefore, our research aims at understanding the change processes and how to overcome problems and obstacles to optimize the programmes. As this chapter was written, DZLM is in the third funding period (2019–2020) with the aim of establishing it as a permanent nationwide operating institute for research and development in the field of mathematics teacher professionalization.

All these aspects could have been illustrated by only one course. However, in order to illustrate how these principles could be interpreted in different courses and because of the history of what has become this chapter (a joint presentation at the ERME conference), we decided to discuss both courses from a design point of view and a research point of view.

The courses chosen are very similar in nature, they address upper secondary teachers who face a new curriculum with mandatory use of digital tools, they both had the same structure – four whole days spread over several months, with working phases in between every two meetings – and were situated in the same educational context, the federal state of North Rhine-Westfalia. Course 1 focused on the use of digital tools (in various domains, including probability and statistics), course 2 on probability and statistics (including the use of digital tools).

Due to limitations of space, we have decided the following:

- The design principles for teaching and learning process will be illustrated by course 1 only.
- The theory-based design principles for the content will be illustrated by course 2 only.
- To present some results of the accompanying research related to both courses.

The research goals and methods differ in both courses to a certain extent due to the early stage of DZLM's development, where different research strategies were practised, which later evolved into a new systematized research agenda (Prediger, Leuders & Rösken-Winter, 2019).

Before we start with the two courses, we will discuss the context of the courses and the principles of the DZLM.

2. Professional development in mathematics education in Germany and the DZLM approach

2.1 General situation of PD courses

Although research findings gave evidence that effective professional development should be realized in long-term settings with more than one face-to-face meeting, these conclusions from research did not yet lead to be realized in the practice of teacher education in Germany.

To get an idea of why these change processes are difficult, it is necessary to briefly describe the educational structure in Germany. Teacher education in Germany is mainly structured in three phases. The first phase at university takes 3.5 to 5 years, depending on students' aims to become a primary or secondary teacher. Although there are standards for teacher education published by the Society of Didactics of Mathematics together with the main teacher association for mathematics (DMV, GDM & MNU, 2008), these are not compulsory but just recommendations. There is still a big variety of how to conceptualize the education at university as the official guidelines are very vague and allow still many ways of how to decide and realize the content of the education. This is the same for the second part of teacher education, which happens over 18 months at special centres for pre-service education, run by the regional school administration. During this phase, the future teachers have to teach at a school, partly supervised and assessed, and partly in their own responsibility. In-service education is not compulsory and is offered under the authority of the school administration and by free providers (such as teacher networks, universities, teachers' association). As in many other countries, there are currently no standards or guidelines for professional development, and facilitators do not receive specific education to be prepared for this job. They are mainly qualified teachers who are denominated by their governments to act as trainers and facilitators.

2.2 The DZLM approach

The DZLM (German Center for Mathematics Teacher Education) was launched in 2011 with the aim to support and pursue the existing programmes and structures for continuous professional development (CPD) nationwide, network all reforms in this field and develop new research-based exemplary courses. DZLM is a joint endeavour of different researchers from different universities, collaborating with representatives from school administration of all 16 federal states and teachers from school practice. DZLM follows a design-based research paradigm (van den Akker, Gravemeijer, McKenney & Nieveen, 2006) when designing and researching programmes for different target groups of teachers and topics. DZLM offers qualification of facilitators, in-service-teacher education and out-of-field teaching and acts as a network platform for information and exchange.

One important issue for the DZLM was to install a competence framework for PD courses as an important orientation for teachers and facilitators to address the different areas of relevant content (see Figure 10.1, cf. Lipowsky & Rzejak, 2015; Garet, Porter, Desimone, Birman & Yoon, 2001).

This framework was the framework of principles for content design that has to be selected from and concretized in each PD course (see course 2 as an example).

Beside the competence framework, DZLM has established design principles for the teaching and learning processes for all contents as guidelines for designing and analyzing CPD courses. This has been done in a cooperative process of all DZLM researchers reviewing the current state of research in the field. Based on this process, six design principles have been generated to provide criteria of efficient teachers' professionalization. They were based in the research literature of effective PD (for further details, see Barzel & Selter, 2015). We have organized

Professional Knowledge			Beliefs	
Subject Specific		General		
Mathematical Content Knowledge	Pedagogical Content Knowledge	Pedagogical Knowledge	Mathematical-related beliefs	Self-related beliefs
Mathematics from a broader perspective	Content-related (diagnostics, tasks, approaches to content)	Knowledge of social and academics education	Mathematics as science	Self-efficacy
				Willingness to cooperate
		Methodology	Mathematics teaching and learning	Identity
School-based mathematical knowledge (according to national standards)	Teaching-related (educational standards, teaching and learning, students' achievement)	Communication and interaction	Interest in mathematics	Job satisfaction
		Heterogeneity		Willingness to innovate
Technical Skills		Pedagogical Content Knowledge for Providing CPD			
Handling of web-based opportunities	Technology-related beliefs	Design of CPD courses	Support of school development	Management of CPD	Reliefs related to CPD
Digital learning platforms	Interest in technology	Didactics of adult education	Schools as learning organizations	Systematic networking and cooperation	Self-efficacy
E-learning	Media-related self-efficacy	Design principles for effective CPD	Consulting and coaching	Organization of CPD courses	Cooperation
Online communication			Professional learning communities		Relevance of design principles

FIGURE 10.1 DZLM competence framework for PD courses.

the research results into six keywords that can also be useful for communicating design principles to facilitators:

- *Competence-orientation*: crucial for effects and efficacy of professionalization is the clear focus on content to improve and deepen teachers' knowledge, and performance in teaching
  (Garet et al., 2001; Timperley, Wilson, Barrar & Fung, 2007)

- *Participant-orientation*: centring on the heterogeneous and individual prerequisites of participants; moreover, participants get actively involved in the PD unit instead of pursuing a simple input-orientation
  (Clarke, 1994; Krainer, 2003)

- *Stimulation of cooperation*: motivating participants to work cooperatively, especially between and after the face-to-face phases; ideally sustainable professional learning communities are initiated
  (Krainer, 2003; Bonsen & Hübner, 2012)

- *Case-relatedness*: using cases such as videos of teaching or students' documentation, which are relevant for the school practice, to enable new perspectives and to realize further dimensions of teaching effects
  (Borko, 2004; Timperley et al., 2007; Lipowsky & Rzejak, 2015)

- *Diverse instruction formats*: during PD courses, it is important to realize a mixture of different formats (like lectures, individual and collaborative work); also, phases of attendance, self-study and e-learning should alternate
  (Deci & Ryan, 2000; Lipowsky & Rzejak, 2015).

- *Fostering reflection*: continuously encouraging participants to reflect on their conceptions, attitudes and practices

(Deci & Ryan, 2000; Putnam & Borko, 2000; Schoen, 1983)

Taking these principles seriously naturally yields to the necessity to realize CPD initiatives in long-term formats as well (Rösken-Winter, Schüler, Stahnke & Blömeke, 2015; Fishman, Penuel, Allen, Cheng & Sabelli, 2013).

The DZLM provides an organizational and supporting structure to accompany the programmes with associated research. One main question is to focus on the effects of CPD programmes. There is a consensus in literature that the effects of CPD occur on different levels, and that just the number of levels varies: Whereas Guskey (2000) defines five levels of effects, Lipowsky and Rzejak (2015) distinguish four levels of effects: level 1, *participant's reactions*; level 2, *participant's beliefs and professional knowledge*; level 3, *participant's use of new knowledge and skill in the classroom*; and level 4, *student learning outcomes*. Guskey's (2000) additional level describes "Organization Support and Change" and is positioned between the second and third levels in the aforementioned hierarchy. Guskey's extra level specifies whole school changes as a result of CPD. Since DZLM focuses less on whole schools but more on individual teachers, the orientation is on the four-level variant.

2.3 Context of the two PD courses

The following two examples of PD courses illustrate the work of the DZLM. Both examples are from North Rhine-Westphalia (NRW). It is the biggest federal state in Germany in terms of numbers of inhabitants (18 million of 82 million in the whole of Germany). The federal states are responsible for any educational issues. The nationwide standards in mathematics (KMK, 2012) serve as a recommendation, but most of the curricula in the federal states follow these standards. In previous years, two main innovations for upper secondary level and the final centralized examination (Abitur) have been brought up in NRW. It is, on the one hand, the introduction of graphic calculators (GC) as compulsory tools (by decree in 2012) in classrooms and examinations. On the other hand, the new state curricula in NRW (2014) fixed stochastics (probability and statistics) as an obligatory topic for all students (6 months' teaching of stochastics in all mathematics classrooms). The main argument for the introduction of the GC was to support a deeper understanding of mathematics by interactive visualization, relief from routine calculations and routine analyses of data and support of modelling with more realistic examples. Regarding stochastics, in particular, the use of the GC for simulations is suggested.

For both topics – an introduction on using and teaching with GCs and on teaching stochastics – the DZLM has collaborated strongly with the educational administration in NRW and realized two PD courses: "GC compact" and "Stochastics Compact". In the following, we present both courses to illustrate the work of the DZLM.

Teacher professional development is essential to further develop mathematics teaching and is still a current issue (Borko, 2004; Sztajn, Borko & Smith, 2017). In recent years, a shift can be stated in better conceptualizing and grounding professional development by means of research. It is no more aimed at eliminating shortcomings, but the development goes more into the direction of a continuous process of professionalization (Rösken-Winter & Szczesny, 2017). That is why the duration and formats of professional development should change from single short courses to courses consisting a mixture of several face-to-face meetings, as well as blended learning phases for supporting teachers (Fishman et al., 2013).

2.4 Common design aspects related to digital tools

Intensive professional development is especially needed to support teachers to involve digital tools in their teaching, as this is a big challenge requiring a rethinking of task formats as well of teaching routines within the orchestration of the classroom. We use the term "digital tools" to describe mathematical software such as spreadsheets, computer algebra systems (CAS), or statistical tools. These tools can be available on various platforms such as desktop computers, tablets, or handheld calculators. Pierce and Stacey (2010) have used the umbrella term "Mathematics Analysis Software" (MAS) to describe this kind of software, housed in a computer or a calculator, which they declare as cognitive tools, "in that they facilitate the technical dimension of mathematical activity and allow the user to take action on mathematical objects or representations of those objects" (Pierce & Stacey, 2010, p. 1). Considerable research over the past few decades, including reviews, has pointed out that MAS in mathematics education can be used to enrich teaching and learning (e.g. Blume & Heid, 2008; Heid & Blume, 2008; Barzel, 2012; Biehler, Ben-Zvi, Bakker & Makar, 2013; Drijvers et al., 2016). For example, these tools can facilitate constructivist teaching approaches like discovery learning by offering the opportunity to explore mathematical connections. In addition, digital tools can enhance conceptual understanding of specific content by providing easy access to multiple, linked and dynamic representations (Penglase & Arnold, 1996; Burrill et al., 2002).

Despite these results and recommendations by researchers, there is a "widely perceived quantitative gap and qualitative gap between the reality of teachers' use of ICT and the potential for ICT suggested by research and policy" (Bretscher, 2014, p. 43). Factors which are discussed as reasons for the reluctance of using MAS by teachers are external factors like time constraints, resources and school culture, but much more important is the teacher him- or herself not changing teaching routines to integrate MAS.

For Germany, a current representative survey of more than 1000 teachers confirmed this fact by stating that STEM teachers do not assume their pioneering role to include technology in teaching, although literature offers a plethora of relevant teaching examples (Lorenz et al., 2017). The point that technology is nearly not used to enhance content- and process-related activities is even more crucial

(Lorenz et al., 2017). This is one important reason why DZLM offers a two-folded programme to support secondary teachers to integrate MAS in their teaching.

3. Design Principles and accompanying research – realized in the PD course "graphic calculators compact"

3.1 Design principles as realized in the course "graphic calculators compact"

The DZLM together with the Ministry of Education in NRW was in charge of developing, delivering and evaluating the long-term professional development (PD) course to integrate graphic calculators (GC) in mathematics classrooms. The project is situated in the context that applying graphic calculators is compulsory in upper secondary level teaching, and in the final centralized exam (called "Abitur") since the beginning of 2014. The design of the course was realized in different design cycles; the first cycle can be characterized as a strong collaboration within a group of teachers, researchers and one person from the school administration. The course was realized in 2014–2015. It consists of four 1-day modules (8 hours each) over a half year with phases of own experiences and elements of blended learning in between (mainly networking to exchange materials).

The DZLM design principles served as the main guideline for the design from the beginning.

Competence-orientation

The PD course covers different dimensions of teachers' competencies, which can be summarized in four main goals. The teachers should be able to use a tool in a flexible way, to design tasks integrating the technology, to organize the classroom in a technology-based environment and to develop appropriate formats and tasks for assessment with the graphic calculator tool according to existing literature (e.g. Barzel, 2012, which is an international review on literature in the field). The four modules were dedicated to these four goals. The first module was about introduction into working with GCs, the second one about designing tasks by integrating the use of GCs, the third one focused on classroom organization in a technology-based environment and in the final one, we worked on assessment issues. The concrete design of the single modules was based on research results. From the beginning of the course, we highlighted relevant subject matter aspects when teaching functions and derivatives integrating technology. For example, we pointed out the importance of developing concept images (Tall & Vinner, 1981; Bingolbali & Monaghan, 2008) and "Grundvorstellungen" (vom Hofe & Blum, 2016) of functions and derivatives and offered tasks to initiate a fluent use and change between mathematical representations (Duval, 2002). Besides these basic aspects, systematic evidence is presented on typical student errors, preconceptions and misconceptions in the field of functions (Swan, 1985; Hadjidemetriou & Williams, 2002; Barzel & Ganter, 2010). Additionally, we always explicated the role of technology as well as possible advantages and burdens when using technology (Barzel, 2012). All these goals are made transparent for all participants, thus enabling teachers to clearly see the relation to their own teaching practice and to increase their motivation while attending the course. The task to introduce the technical facilities during the first module was the task "power flower" shown in Figure 10.2 (Barzel & Möller, 2001). This task served as an investigative open task as well as an example of meaningful tasks when integrating technology and offered opportunities to reflect on the value of technology concerning the aforementioned aspects of pedagogical content knowledge. Module 2 offered a sample of modelling tasks in the field of mathematical topics for the upper secondary level, also including opportunities for data logging. Module 3 picked up the power flower task (Figure 10.2) to discuss classroom organizations and the point that technology can be used either to introduce a new topic (e.g. here power functions) or to deepen knowledge during a final phase of an exercise. Module 4 focused on exam situations. Current examination tasks were analysed as to whether the use of graphics is necessary, supportive, neutral, or forbidden. Another perspective of reflecting the tasks was the role of the technology for which the graphics are used: for discovery learning, for conceptualizing, for enabling individual approaches, for taking over procedures, or for controlling. This categorization is also suggested by the current German standards for mathematics at the upper secondary level (KMK, 2012)

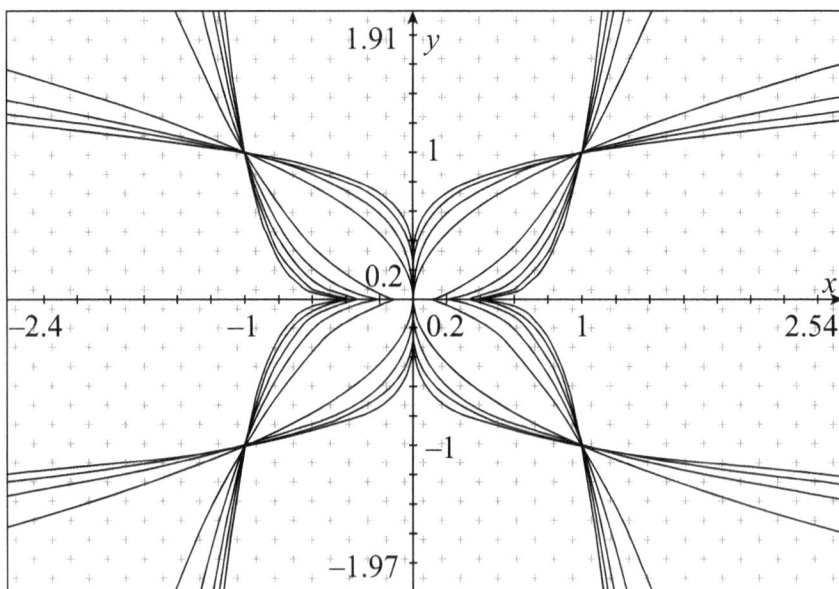

FIGURE 10.2 Task to get familiar with the GC: "Create this picture on your screen!"

Participant-orientation

The participant-orientation combines two challenges: taking up heterogeneous competences and conditions, and fostering participants' self-responsibility.

Initially, a preliminary questionnaire regarding the teachers' conditions, expectations and needs with respect to content and didactical issues can help to adapt the course to the specific target group. All tasks during the course are created to be used in the classroom with students as well. Accompanying material and information about the tasks show possible solutions, typical errors and misconceptions, an idea of how and where to integrate the tasks in the learning process and the relevant role of technology. To foster self-efficacy and self-responsibilities, it is important to include a lot of opportunities which activate the participants – such as working on tasks in pairs and small groups and initiating discussions and reflections about the material. Furthermore, at the end of each course, participants were actively involved in providing recommendations for content and methodology that should be included in the following meetings.

Stimulation of cooperation

Aiming at sustainable cooperation processes, we tried to stimulate professional learning communities (PLC) with teachers from one school or neighbouring schools during the first face-to-face meeting. This stimulation was accompanied by a short input about the importance and power of intense collaboration in PLCs. However, because there was only one teacher from each school, the initiation of the PLCs was not successful.

Case-relatedness

All modules relate to practical experiences by discussing ideas based on specific cases from classrooms. On the one side, we brought cases into the courses such as specific student results and examples. On the other side, we asked the participants to bring own cases from their classrooms to provide both a starting point for discussion and an impulse for reflection. Figure 10.3 gives an impression of how such cases are used – here to discuss the challenge how students' documentation and language should look like when computer algebra is used. Here, we used the recommendations of Schacht (2017) to distinguish that the use of technical expressions in the documentations can be allowed when learning to become familiar with the technology, but that consolidated mathematical language must be used at the end of the learning process.

Various instruction formats

To ensure active participation and the experience of self-efficacy, various instruction formats are used throughout all face-to-face meetings. The whole PD course includes phases of attendance, self-study and e-learning to initiate cycles of input, learning, practical try-outs and reflections.

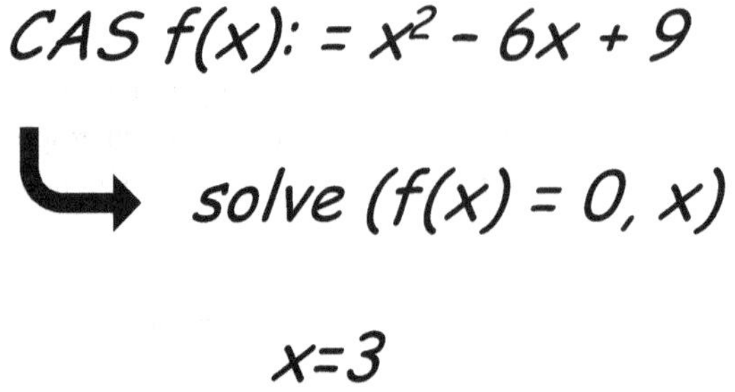

$$CAS\ f(x){:} = x^2 - 6x + 9$$

$$\hookrightarrow solve\ (f(x) = 0,\ x)$$

$$x=3$$

FIGURE 10.3   Is this documentation acceptable or not?

Fostering reflection

Participants are inspired to become "reflective practitioners" (Schoen, 1983) by stimulating cooperative reflection as well as self-reflection continuously with respect to tasks, students' solutions and thinking, scenarios of classrooms and on own conceptions, attitudes, beliefs and teaching routines and practices. Participants were encouraged to think deeply about the possible transfer of the teaching material into their own classrooms, as well as the impact on their own teaching style.

3.2 Research of the PD course

The whole PD course was realized in 2014–2015 for three groups of teachers at different locations in NRW with about 100 participants. The current version of the course material is enlarged now on digital tools instead of graphic calculators and it is published under Creative Commons license on the national DZLM server:

> https://dzlm.de/angebote/angebotssuche/field_angebotstyp/fortbildungs materialien-432

The associated research project focused on (a) the effectiveness of the PD course as well as on (b) conditions and criteria that have to be considered when designing a PD with respect to graphing technology. In this paper, we focus on (a) the research on the effectiveness of the PD course. Regarding (b), we refer to the publication of Thurm, Klinger, Barzel and Rögler (2017), Thurm (2017) and Klinger (2018). The study regarding (a) was structured according to the different levels on which effects of PD courses can occur (e.g. Lipowsky & Rzejak, 2015).

Level 1: participant's reactions

To assess participants' reaction to each module, we administered a questionnaire after each module that asked participants to what extent the design principles were realized. In addition, the participants rated their overall experience at the end of the PD course.

Level 2: participant's beliefs and professional knowledge

Effects of the PD course at level 2 were investigated with a focus on teacher beliefs. We used a questionnaire regarding teachers' epistemological beliefs about the nature of mathematics and beliefs about the learning of mathematics (Blömeke, Hsieh, Kaiser & Schmidt, 2014). In addition, we collected data on the self-efficacy beliefs of teachers with respect to developing and selecting appropriate tasks and with respect to designing lessons when teaching with technology. Finally, beliefs regarding the value of technology for teaching mathematics were gathered by a questionnaire (Thurm, 2017; Thurm & Barzel, 2020).

Level 3: participant's use of new knowledge and skills in the classroom

Effects of the PD course at the classroom level were evaluated by a questionnaire capturing the self-reported frequency of technology use (Thurm, 2017; Thurm & Barzel, 2020). We did not focus on quality since it is known that survey data is well suited for describing quantity but not as suitable for describing the quality of instruction.

Level 4: student learning outcomes

On the student level, we designed two tests assessing, for example, the concept image of functions and the ability to switch between multiple functional representations (Klinger, 2018).

We chose a classical pre-post-test control group design and administered the questionnaires to teachers participating in the PD course and to a control group of teachers not participating in the PD course. Figure 10.4 provides an overview of the whole project.

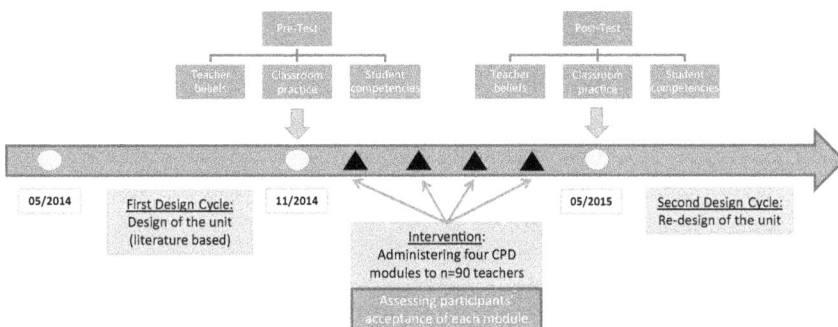

FIGURE 10.4 Overview of the research project.

Out of 90 participants of the PD course, 40 volunteered to take part in our research. The control group consisted of 89 teachers, who were enlisted by a circular letter and an associated website. To ensure that experimental and control groups were equivalent with respect to the attributes under scrutiny, we used the propensity score matching method (Rosenbaum & Rubin, 1985) to identify 38 teachers out of the 89 initial control group teachers that were most similar to the experimental group.

Results on level 1: participant's reactions

The analysis of the questionnaire (see Table 10.1) that focused on the acceptance of the PD course showed that teachers rated the course highly.

In particular, the teachers emphasized the high relevance of the topics for their daily classroom work. Hence, it can be said that the PD course had a positive effect on level 1.

Results on level 2

With regard to self-efficacy beliefs, it could be shown that these beliefs increased in the experimental as well as in the control group. However, the size of the gains was comparable in both the experimental and the control group and therefore we could not identify an effect of the PD course regarding self-efficacy beliefs. Effects were also missing when looking at epistemological beliefs. This result might be due to the fact, that teacher in the control and the experimental group showed very strong constructivist beliefs. The scales, therefore, may not be suitable to detect changes on a subtler level.

When looking at beliefs referring to teaching with technology, there were significant effects of the PD course (Thurm, 2020, p. 266f). As can be seen from Figure 10.5, teachers in the control group showed a decline in the belief that technology can support discovery learning (see Figure 10.5a) and multiple representations (see Figure 10.5b). In addition, the control group was more convinced that technology integration is time-consuming (see Figure 10.5c) and that technology should be used only at the end of the learning process (see Figure 10.5f). Contrary to this, teachers in the experimental group did not show such a negative development. Instead, beliefs remained unchanged or improved towards beliefs supporting the use of technology.

TABLE 10.1 Rating of the PD course.

How do you rate the PD course	M	SD
I gained mathematical knowledge.	4.30	1.60
I gained pedagogical content knowledge.	4.19	1.52
The PD material helped me in my daily teaching.	4.32	1.29
The content of the PD course was relevant for my daily teaching.	4.78	1.25
I would recommend the PD course.	4.78	1.42

Note: Response format: 1 = do not agree, 6 = fully agree.

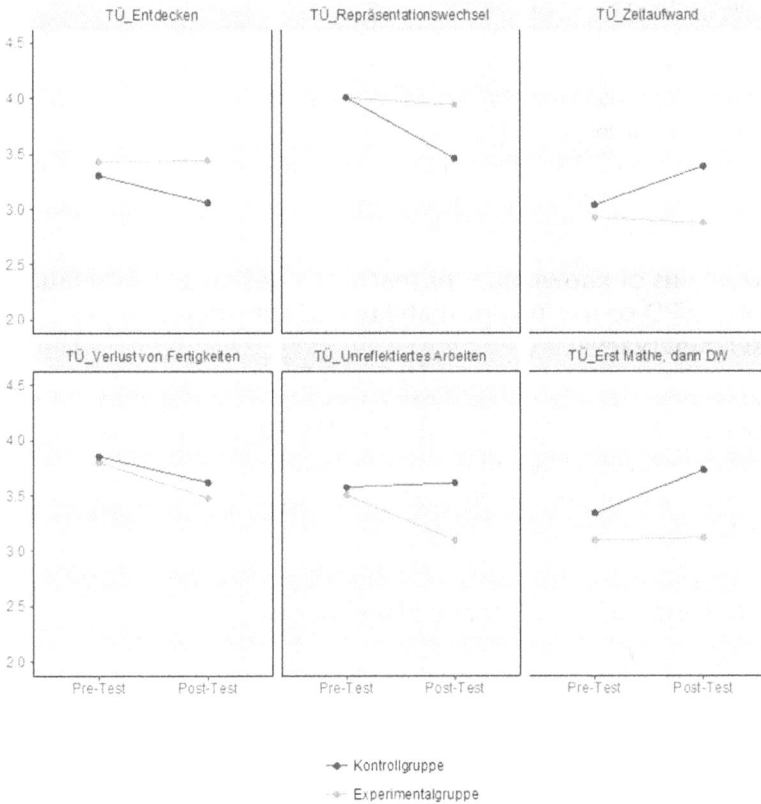

FIGURE 10.5 Effects of the PD course on teachers' beliefs about the value of technology.

These results must be interpreted in light of the fact that the majority of teachers in this study had just started to teach with technology. Therefore, the beliefs of these teachers regarding the use of technology were still very susceptible to change through practical teaching experiences. Hence it seems plausible to infer that teachers in the control group had substantial negative experiences when integrating technology, which could be due to the challenges and complexity inherent in the process of integrating technology into daily teaching. In contrast, teachers in the control group were apparently better prepared to deal with these challenges due to the support of the PD course. This points to the direction that PD courses are especially important for teachers that just begin to integrate digital technology.

Results on level 3

When looking at the frequency of technology use, we noticed that teachers in the experimental group used the GC more than teachers in the control group. However, this effect was not significant. This might reflect that it takes time for teachers to change their routines, and that change is more an evolutionary process.

Results on level 4

On this level, we could not identify any effect of the PD course. Pupils of teachers in the experimental group did not perform better on the test compared to pupils in the control group. However, this result is not surprising given the limited effects on level 3 and the fact that effects on level 4 take the most time to unfold (Guskey, 2000).

4. Domains of knowledge for teachers – design and evaluation of the PD course on "probability and statistics at upper secondary level"

In this section, we will focus on the design of a PD course from the perspective of the facets of teachers' knowledge that we addressed. The course also considered the design principles of the previous section, but as we explained in the introductory section, we will not make this explicit but focus on content design and related research.

4.1 Context and overall design of the course

We will illustrate how the content of a PD course was selected and the design was developed based on several circles of implementation and further elaboration. Our focus is on a PD course for upper secondary Gymnasium teachers, which we named "Stochastics Compact". Stochastics is used in Germany for the combination of probability and statistics. The course lasted 4 months, with four and later five 1-day meetings. We started in 2013 in the state of North Rhine-Westphalia and improved the material in 5 revision cycles. A total number of about 500 teachers have participated in the various versions of the course.

Versions 1.0 and 2.0 of the course were designed by a DZLM team that consisted of teachers and young and senior researchers, including the second author of this chapter. From version 3.0 onwards we entered into a collaborative project with three facilitators from the federal state of Thuringia and five facilitators from the region of Arnsberg (3.6 million inhabitants) in North Rhine-Westphalia, with whom we developed new versions of the material and jointly used the material in our courses. The collaborative development, implementation and reflection aimed at improving the materials and at qualifying the three-plus-five facilitators at the same time; we call them "project facilitators" in contrast to the other facilitators that may use the material but who were not part of the development team. All eight facilitators were experienced teachers who have been active as facilitators for many years; however, long-term PD courses such as "Stochastics Compact" were new for them. The fact that we brought version 2.0 of the course into the collaboration was a good starting point.

In Arnsberg, the regional administration supported a collaboration that lasted more than three years and three development cycles. The materials reached a final stage (version 4.0) in October 2017 and were ready for use by all mathematics facilitators of the Arnsberg region. We have published a further elaborated version of a part of the material under Creative Commons license on the national DZLM server (https://dzlm.de/angebote/angebotssuche/field_angebotstyp/fortbildungs materialien-432). In 2018, we started a collaboration with the federal state of Brandenburg, where we are qualifying about 20 facilitators.

4.2 Collaborative design of the PD course – cooperation between DZLM team and regional facilitators

The factors that finally influenced the design of the materials are multifaceted, as is shown in Figure 10.6.

The picture (Figure 10.6) depicts some tensions between different *views* of the needs of mathematics teachers. As the facilitators and the regional PD admin- istrators were part of the design team, the final design was not just based on DZLM views but a result of a complex negotiation process. The DZLM team is rooted in the knowledge base and research and development tradition of stochas- tics education. The new curricula only partly reflect major suggestions and ideas from this tradition. Our course is compatible with the new curricula but tries to influence how these new curricula are interpreted and realized in the classrooms. The syllabus of the curriculum allows options for school-based developments

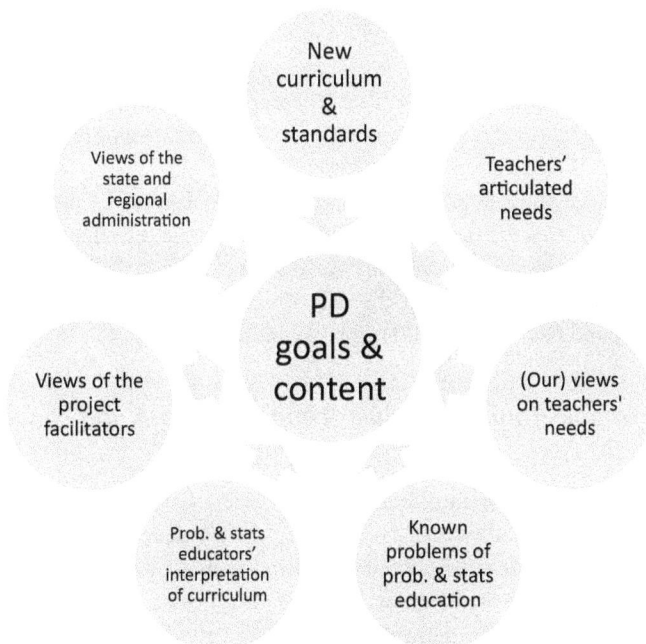

FIGURE 10.6 Influencing factors for the PD material.

Subject matter didactics: "Fundamental ideas"	Problems of understanding; research on students	Own Research on classroom experiments; practical experiences
Connecting data and chance	Probability concept and relative frequency; laws of large numbers	Grade 9–11
Stochastic independence	Independence as an implicit assumption, and sometimes an inadequate one	Pre-service teacher education
Bayesian reasoning, "natural frequencies approach"	Confounding conditional probabilities, e.g. AIDS tests; percentages in media	Grade 9
Binomial distribution as a model	Naïve modelling without assumption checking	Pre-service teacher education
Hypothesis testing	Misinterpretations of P-values and of results of hypothesis testing in general	Grade 12

FIGURE 10.7   Subject matter components of the course and related research.

and variation, and we intend to use this free space. We address teachers as independent personalities that we support in developing their own view of stochastics and stochastics education, and we do not treat them just as curriculum implementers. We base the selection of PD content on analyses of difficulties of students and teachers and on a view of "fundamental ideas" for teaching stochastics at the upper secondary level (Burrill & Biehler, 2011; Biehler & Eichler, 2015). We build on insights on how technology can be used to support students' learning in stochastics (Biehler et al., 2013) (see Figure 10.7).

Moreover, we suggest teaching approaches and material that we had used in university courses for future teachers or that we had tested and scientifically evaluated in experimental classrooms: for instance, Meyfarth (2006) on hypothesis testing, Prömmel (2013) on the use of simulations, and Wassner, Biehler, Schweynoch and Martignon (2004) for Bayesian reasoning.

4.3 An example from the module "connecting data and chance"

For illustrative purposes, we describe an example from the topic "connecting data and chance" (from the first module of the course).

We start with the following "landmark" activity that we are suggesting as a classroom activity when introducing stochastics at the upper secondary level: as a challenging problem for students which will also show the power of computer-based simulations for solving problems in probability.

The task (Figure 10.8) has the same mathematical structure as the "maternity ward" problem that was used by the psychologists Daniel Kahneman and Amos

Tversky in their research on wrong human intuitions on the role of large numbers (Tversky & Kahneman, 1971; see later in this section). Activities in the PD course include: guessing intuitively, some initial discussion about reasons for the choices and use of simulation to decide the question (estimate the probabilities to pass the test just by guessing). In all our courses, all three answers were initially chosen by at least some of the teachers, this was always used for stimulating interesting and lively discussions.

We start with simulation by hand (with a coin) where the small sample size usually does not provide a clear answer, and then we move to computer-based simulation (with a GC) to get more precise and reliable results. Estimating the passing probability will be supplemented by visualizing the whole distribution of "proportion of correctly answered questions" (see Figure 10.9 right). This is the basis

Students can choose between two multiple choice tests with two choices in each question (one choice is correct)

Test 1: 10 questions

Test 2: 20 questions

A test is passed if at least 60% of the questions are correctly answered.

If a student just guesses: Which test is easier to pass?

O Test 1   O Test 2   O Equal chances

FIGURE 10.8   The 10–20-Test problem.

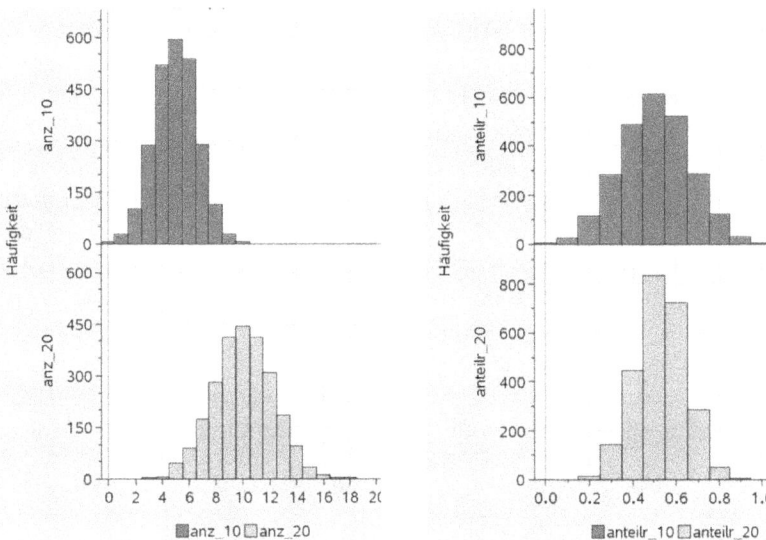

FIGURE 10.9   Simulation and visualization of the distributions with the TI Nspire.

for integrating the results into an elaborated intuitive view of how the distribution of relative frequency changes with increasing sample size.

Some teachers can relate the picture on the right side of Figure 10.9 to their intuition that the relative frequency tends to be closer to the expected value of 0.5 when the sample size is larger. This stems from intuitions about the law of large numbers, although most of our teachers have never seen such a display as the law of large number is often only visualized as a trajectory, where the relative frequency "approaches" the theoretical probability. The left side shows the simulated distribution of the number of successes, where the spread is increasing. We support our teachers in relating this to their previous knowledge. The number of successes of guesses during the testing can theoretically be modelled as random variables X_n with a binomial distribution with expected value at $n \cdot p$, that is at 5 respectively 10, and a standard deviation of $\sigma = \sqrt{n \cdot 0.5 \cdot 0.5}$, which increases with n. The right-hand side of Figure 10.9 is a simulation of the random variable $Y_n = X_n / n$, whose standard deviation is $\dfrac{\sigma}{n} = \dfrac{\sqrt{0.5 \cdot 0.5}}{\sqrt{n}}$.

A next step is to widen the question to what will happen when we further increase the sample size n. Some teachers know that the middle 95% prediction interval around 0.5 can be theoretically calculated as $\left[0.5 - 1.96 \cdot \dfrac{\sigma}{n} ; 0.5 + 1.96 \cdot \dfrac{\sigma}{n} \right]$, which is roughly $\left[0.5 - \dfrac{1}{\sqrt{n}} ; 0.5 + \dfrac{1}{\sqrt{n}} \right]$, and its width is $\dfrac{2}{\sqrt{n}}$. The so-called normal approximation of the binomial distribution is used for deriving this interval. The specific dependence on the sample size n is also called the "one-over-square root-of-n-law". This knowledge is considered as knowledge "at the mathematical horizon" in the sense of Ball and Bass (2009). This cannot and should not become the topic of instruction at the beginning of the stochastics course, but it is important for teachers' orientation.

We then introduce to our teachers a way for introducing the "one-over-square root-of-n-law" just based on simulations and visualizations by means of "the prediction activity". Based on simulated data, the percentile commands are used to find the middle 95%–interval (Figure 10.10, left side) and the GC is then further used to explore how the width of this interval depends on the sample size n (see Figure 10.10, right side).

It is claimed (without proof) that this law can be generalized to any p and n. For n repetitions of a random experiment with success probability p, the following inequality holds with 95% probability for the relative frequencies

$$f_n : |p - f_n| \leq \frac{1}{\sqrt{n}} \quad \text{(95%-prediction intervals)}.$$ We argue that this knowledge is

important for students when they have to relate data and chance: instead of a vague idea that the relative frequency tends to approach the probability p with

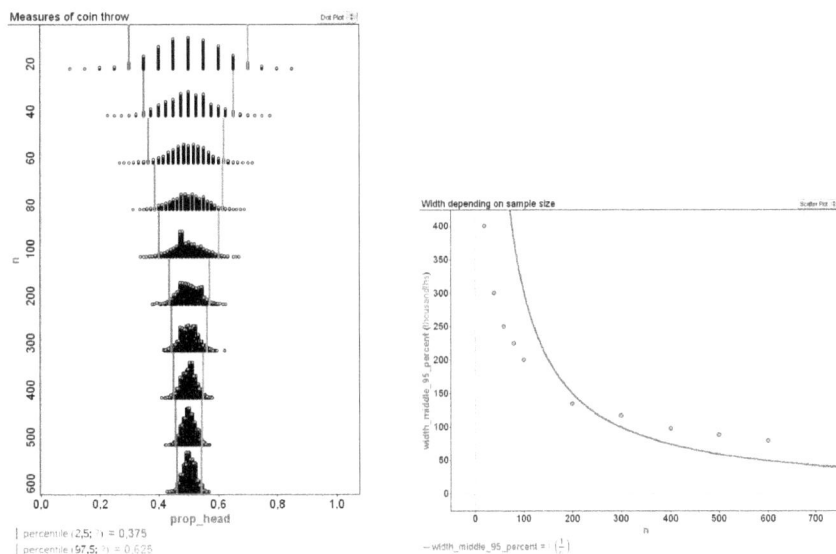

FIGURE 10.10   Left side: empirical 95%-prediction intervals. Right side: Trying to fit a curve to the width of the middle 95%.

Note: Functions such as $\frac{k}{n}$ do not work for any k; $\frac{k}{\sqrt{n}}$ fits well for k = 2.

increasing n, an interval can be provided in which we can expect the relative frequency with 95% certainty.

We also argue for introducing the inverse statement (with some horizon knowledge on confidence intervals that we cannot elaborate on in this chapter). If p is unknown, we usually (with 95 % probability) will observe a relative frequency f_n, a value which cannot be "far" from the true probability p: $\left|p - f_n\right| \le \dfrac{1}{\sqrt{n}}$. The practical value for students is that if they observe f_n, they can provide a so-called intuitive 95%-confidence interval for p, namely $\left[f_n - \dfrac{1}{\sqrt{n}}, f_n + \dfrac{1}{\sqrt{n}}\right]$.

This knowledge is not obligatory in the German syllabus, but we argue that a sound dealing with simulations in the classroom requires knowledge about how precisely the unknown probability can be estimated from the relative frequency and how certain this estimation is.

We suggest that teachers at least communicate a rule of thumb table to their students containing interval widths for "standard" sample sizes (Table 10.2).

4.4 Addressed facets of teachers' knowledge and beliefs

We base our course on models of teachers' knowledge such as that of Hill, Loewenberg Ball and Schilling (2008, p. 377). The authors distinguish Common

TABLE 10.2 Prediction and confidence intervals for standard sample sizes.

Sample size	Radius of 95%-prediction interval / intuitive confidence interval
50	± 0.14
100	± 0.10
1.000	± 0.03
10.000	± 0.01

Content Knowledge (CCK), Knowledge at the Mathematical Horizon (HK), Specialized Content Knowledge (SCK), Knowledge of Content and Students (KCS), Knowledge of Content and Teaching (KCT) and Knowledge of Curriculum (KC). This classification, however, leaves open what counts as "knowledge" and what the warrants are for the respective knowledge base. We try to overcome the situation that this knowledge is solely based on opinion and experience, and we introduce results from research in psychology and mathematics education as evidence for our suggestions and claims.

An extension that takes into account the various facets when we include technology was developed in Wassong and Biehler (2010).

We will illustrate only some of the facets that are shown to Figure 10.11. TK (Technological Knowledge) includes basic aspects of using the graphic calculator, TCK (Technological Content Knowledge) includes how to use the GC for

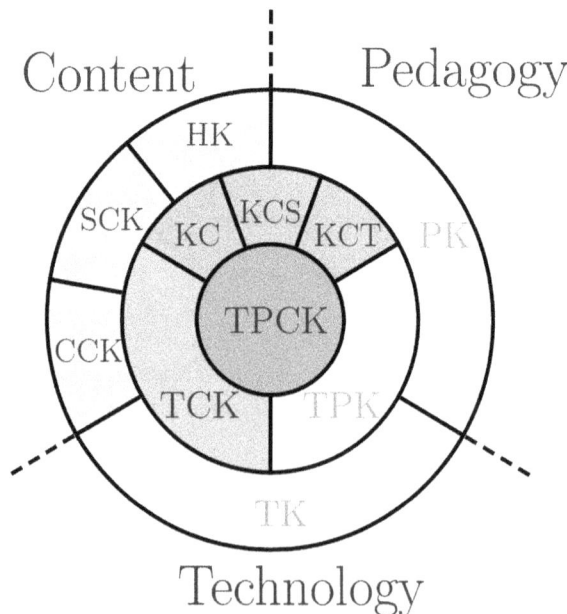

FIGURE 10.11 Extended domain map including facets of knowledge on technology, figure similar to Wassong and Biehler (2010, p. 2).

simulations in stochastics, and TPCK (Technological Pedagogical Content Knowledge) includes how to use the GC so that students can develop a better understanding of the law of large numbers through interactive experiments and simulations. KCT includes the suggested activities (10-20-Test, prediction activity) and which representations to use for the simulated distributions. We already mentioned the knowledge at the Mathematical Horizon (HK), that is, background knowledge by which teachers can judge whether our suggested simplifications are still an adequate elementarization of genuine mathematical content, and why the topics are important to teach. KCS includes misconceptions concerning the role of sample size. On a practical level, we include a variety of students' answers and reasonings to the 10-20-Test problem to prepare teachers for what can be expected in the classroom. Moreover, the discussion in the PD course itself – where some teachers have the same misconception at the beginning – is also a source for this knowledge. A mixture of KCS and HK is provided by drawing the teachers' attention to psychological studies, which show the insensitivity to sample size of many students and adults, and the need to better teach this for improving individuals' capacity to adequately reason under uncertainty. We quote the "maternity ward problem", which has the same structure as the 10-20-Test problem, from original sources:

> A certain town is served by two hospitals. In the larger hospital about 45 babies are born each day and in the smaller hospital about 15 babies each day. As you know, about 50% of all babies are boys. The exact percentage of baby boys, however, varies from day to day. Sometimes it may be higher than 50%, sometimes lower. . . . Which hospital do you think is more likely to find on one day that more than 60% of babies born were boys?
>
> *(Sedlmeier & Gigerenzer, 1997, p. 36, based on research by*
> *Kahneman & Tversky, 1972)*

We learned, however, that making reference to the psychological literature alone is not always convincing enough for our teachers. So we asked teachers in our course to become researchers themselves in that they should give the 10-20-Test problem to a selection of their students. Teachers of the 2014 course asked their students ($n = 1163$). The results can be seen in Table 10.3 and convincingly show how widespread these wrong preconceptions are.

In order to give an impression of what can be achieved by teaching, in Table 10.4 we refer to the experimental course of Prömmel (2013, p. 493).

This result resonates well with teachers' experience that even the best teaching will not change all students' minds, but that teaching can be successful for the majority of students.

4.5 Research on teachers' knowledge before and after the course

What have teachers learned during the course? The rules set up by the administration for such courses did not make it possible to administer a knowledge test

TABLE 10.3 Students' response to the 10-20-Test.

%	Grade 5–9	Grade 10–12
Test 1	16	18
Test 2	41	27
Equal chance	43	55

Note: n = 1163; convenience sample.

TABLE 10.4 Students' response to the maternity ward problem before and after teaching.

%	pre	post	Pre: adequate reasoning	Post: adequate reasoning
Test 1 is correct	26	77	18	59

before and after the course. Therefore we used a questionnaire after the course and asked the teachers to subjectively assess their knowledge gain throughout the course (Nieszporek & Biehler, 2017; Lam & Bengo, 2003). This is more valid than a pre-post-design, because teachers may judge their knowledge high in some aspects before the course but recognize only afterwards that this knowledge was more limited. This questionnaire covers various facets of teachers' knowledge, for instance: CK, "I know suitable tasks for school lessons in the context of the topic XXX?"; KCS and KCT, "I am capable of processing and visualizing the content of the topic XXX understandable way for students?" or "I am capable of recognizing and reacting towards misunderstandings and pupils' faulty reasoning regarding topic XXX"; or TPCK, "I am capable of using the GC in context of the topic XXX in school lessons in a didactically advantageous way".

We asked the teachers to express the level of agreement by providing school grades from 1 to 6 (1 = excellent, 2 = good, 3 = satisfactory, 4 = pass, 5 = poor, 6 = fail). Table 10.5 shows the results related to the first module of the first series of courses in Arnsberg, while Table 10.6 shows results related to the first module of the second series of courses.

On average, teachers judge their improvement by about one grade with some easily interpretable differences. The gain in content knowledge (1) is about 1.5, whereas the ability to create a new task is only between 0.5 and 0.7 (5).

We created two scales from the 7 items: items 1 and 2 form the "knowledge scale", whereas items 3 to 7 form the "competence scale". With the competence scale we assess the self-efficacy judgements with regard to future teaching, and in this sense, we go one step further on the chain of influence. We have also built the scales to get a more reliable measurement of this construct, because the items are nearer to classroom activities. We checked reliability by calculating Cronbach's alpha for pre-post and for two cycles of the PD course. The 16 values

TABLE 10.5 Pre and post score means for content-specific facets simulations (Sim.) and the $\frac{1}{\sqrt{n}}$-law ($\frac{1}{\sqrt{n}}$), rated via grades from 1 = excellent to 6 = fail.

		1	2	3	4	5	6	7
		I know suitable tasks for school lessons in the context of the topic XXX	I have understood the tasks of topic XXX and can solve and interpret them autonomously.	I am capable of creating new tasks for this topic XXX	I am capable of preparing and visualizing the content of the topic XXX in an understandable way for students.	I am capable of using tasks with an authentic context in an advantageous way for topic XXX	I am capable of using the GC for topic XXX in school lessons in a pedagogically effective way.	I am capable of recognizing and reacting to misunderstandings and pupils' erroneous reasoning in topic XXX.
Sim.	Pre	4.03	3.31	4.00	3.68	3.77	4.20	3.68
	Post	2.47	2.43	3.31	2.57	2.73	3.03	2.60
$\frac{1}{\sqrt{n}}$	Pre	3.97	3.33	3.71	3.63	3.87	4.00	3.53
	Post	2.77	2.60	3.29	2.80	3.00	3.17	2.86

Note: First cycle, sample sizes between 27 and 31. Please be aware that a decrease in the mean stands for an increase in knowledge, as the Germany, school grade 1 is the best grade, and grade 6 the worst.

of Cronbach's alpha are larger than 0.83 in 14 cases; in 2 cases the values are .767 and .794, which is an excellent result.

Several phenomena stand out: the knowledge increase is higher than the competencies' increase, which is what one would have expected. Both show positive results, which is especially important because the course has affected the self-efficacy of the teachers with regard to the comeptencies as well. The increase in simulations is slightly higher than with regard to the $\frac{1}{\sqrt{n}}$-law.

Table 10.7 shows the results of the second cycle with the revised material. The increase with regard to simulations is higher in knowledge and competencies than in cycle 1, starting at about the same level. The increase with regard to the $\frac{1}{\sqrt{n}}$-law is similar to cycle 1, but as we started with a lower level, the final state is lower than in cycle 1.

Beyond the mean improvements, we experience a high variability (see Figures 10.12 and 10.13).

TABLE 10.6 Cycle 1. Mean scores for knowledge and competence scales, separately for pre and post items and the four content-specific facets, 1 = excellent to 6 = fail.

Content-specific facet	Knowledge (mean)	Competencies (mean)
Simulations pre	3.69	3.86
Simulations post	2.45	2.86
$\frac{1}{\sqrt{n}}$-law pre	3.75	3.81
$\frac{1}{\sqrt{n}}$-law post	2.68	3.04

Note: Sample sizes between 27 and 31.

TABLE 10.7 Cycle 2. Mean scores for knowledge and competence scales, separately for pre and post items and the four content-specific facets, 1 = excellent to 6 = fail.

Content-specific facet	Knowledge (mean)	Competencies (mean)
Simulations pre	3.68	3.57
Simulations post	2.07	2.21
$\frac{1}{\sqrt{n}}$-law pre	4.29	4.31
$\frac{1}{\sqrt{n}}$-law post	3.11	3.13

Note: Sample sizes between 27 and 31.

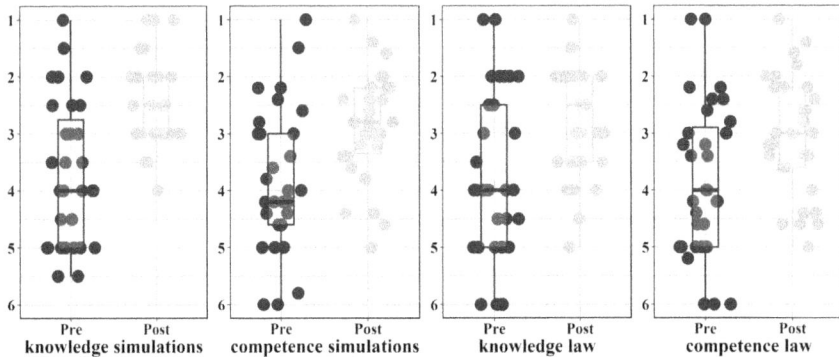

FIGURE 10.12 Cycle 1. Jitter plots and box plots of the two scale pairs knowledge and competence, the content-specific facets simulations and the $\frac{1}{\sqrt{n}}$-law, black: pre scores, grey: post scores (sample sizes between 27 and 31).

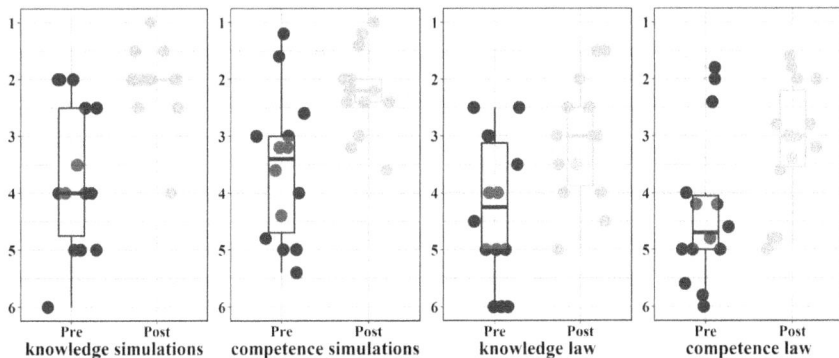

FIGURE 10.13 Cycle 1. Jitter plots and box plots of the two scale pairs knowledge and competence, the content-specific facets simulations and the $\frac{1}{\sqrt{n}}$-law, black: pre scores, grey: post scores (sample sizes between 13 and 14).

Beyond the increase in level, the plots show that in cycle 2 also, the variability was reduced with regard to simulation. With regard to the $\frac{1}{\sqrt{n}}$-law we can identify subgroups in cycle 2 that deviate from the overall pattern.

Future perspectives

Integrating graphing technology into mathematics teaching and teaching probability and statistics were both new challenges for teachers in North Rhine-Westfalia. The research in both courses – the one concerning graphing tools and the other one focusing on probability and statistics – has confirmed the value of long-term PD courses to support teachers when implementing innovations to their teaching. It was remarkable that beliefs referring to teaching with technology showed significant effects through the PD course. Where these beliefs of teachers in the control group declined, the beliefs of teachers in the experimental group remained unchanged or improved towards beliefs supporting the use of technology for discovery learning and the use of multiple representations. The control group was more convinced that technology integration is time-consuming and that technology should be used only at the end of the learning process.

With regard to the course in probability and statistics, it is remarkable that the self-reported gains in knowledge and competence did also reach the self-efficacy judgements with regard to future teaching. Nevertheless, the study also shows the high heterogeneity of teachers before and after the course, which future courses have also to deal with. In more recent versions of the PD course, we offer the teacher more choices of what kind of activities they focus on in the course, considering the different starting points and interests, which may be on technology, deeper content knowledge, or classroom activities.

Both courses are being further developed, published and used in other federal states.

The research on the stochastics course will be part of the PhD project of Ralf Nieszporek, who will also focus on how facilitators shape and implement the jointly developed material. Daniel Thurm's PhD project is about the research on the GC Course according to teachers' beliefs and knowledge (Thurm, 2020; Thurm & Barzel, 2020).

Acknowledgement

Many thanks go to Ralf Nieszporek and Daniel Thurm for providing the data, tables and graphs from the evaluation of the stochastics course (Ralf Nieszporek) and the GC Course (Daniel Thurm).

This chapter is based on Barzel, B., & Biehler, R. (2017). Design principles and domains of knowledge for the professionalization of teachers and facilitators – two examples from the DZLM for upper secondary teachers. In S. Zehetmeier, B. Rösken-Winter, D. Potari & M. Ribeiro (Eds.), *Proceedings of the Third ERME Topic Conference on Mathematics Teaching, Resources and Teacher Professional Development (ETC3, October 5 to 7, 2016)* (pp. 16–34). Berlin: Humboldt-Universität zu Berlin. We have updated and elaborated the text and added results from evaluations of the two courses. Some longer passages of the text are the same.

References

Ball, D. L., & Bass, H. (2009). With an eye on the mathematical horizon: Knowing mathematics for teaching to learners' mathematical futures. In M. Neubrand, et al. (Ed.), *Beiträge zum Mathematikunterricht 2009.* Münster: WTM-Verlag. Retrieved from www.mathematik.uni-dortmund.de/ieem/BzMU/BzMU2009/Beitraege/Hauptvortraege/BALL_Deborah_BASS_Hyman_2009_Horizon.pdf

Barzel, B. (2012). *Computeralgebra im Mathematikunterricht. Ein Mehrwert – aber wann?* Münster: Waxmann.

Barzel, B., & Biehler, R. (2017). Design principles and domains of knowledge for the professionalization of teachers and facilitators – Two examples from the DZLM for upper secondary teachers. In S. Zehetmeier, B. Rösken-Winter, D. Potari, & M. Ribeiro (Eds.), *Proceedings of the Third ERME Topic Conference on Mathematics Teaching, Resources and Teacher Professional Development (ETC3, October 5 to 7, 2016)* (pp. 16–34).

Barzel, B., & Ganter, S. (2010). Experimentell zum Funktionsbegriff. *Praxis der Mathematik in der Schule, 52*(31), 14–19.

Barzel, B., & Möller, R. (2001). About the use of the TI-92 for an open learning approach to power functions. *Zentralblatt für Didaktik der Mathematik, 33*(1), 1–5.

Barzel, B., & Selter, C. (2015). Die DZLM-Gestaltungsprinzipien für Fortbildungen: JMD: Special issue. *Lehrerfortbildung/Multiplikatoren Mathematik – Konzepte und Wirkungsforschung, 36*(2), 259–284. doi:10.1007/s13138-015-0076-y

Biehler, R., Ben-Zvi, D., Bakker, A., & Makar, K. (2013). Technology for enhancing statistical reasoning at the school level. In M. A. Clements, A. J. Bishop, C. Keitel, J. Kilpatrick, & F. K. S. Leung (Eds.), *Third international handbook of mathematics education* (pp. 643–689). New York: Springer.

Biehler, R., & Eichler, A. (2015). Leitidee Daten und Zufall. In W. Blum, S. Vogel, C. Drüke-Noe, & A. Roppelt (Eds.), *Bildungsstandards aktuell: Mathematik für die Sekundarstufe II* (pp. 70–80). Braunschweig: Diesterweg.

Bingolbali, E., & Monaghan, J. (2008). Concept image revisited. *Educational Studies in Mathematics, 68*, 19–35.

Blömeke, S., Hsieh, F.-J., Kaiser, G., & Schmidt, W. H. (2014). *International perspectives on teacher knowledge, beliefs and opportunities to learn.* Dordrecht: Springer.

Blume, G. W., & Heid, M. K. (Eds.). (2008). *Research on technology and the teaching and learning of mathematics: Volume 2 cases and perspectives.* Charlotte, NC: IAP.

Bonsen, M., & Hübner, C. (2012). Unterrichtsentwicklung in Professionellen Lerngemeinschaften. In K.-O. Bauer & J. Logemann (Eds.), *Effektive Bildung* (pp. 55–76). Münster: Waxmann.

Borko, H. (2004). Professional development and teacher learning: Mapping the Terrain. *Educational Researcher, 33*(8), 3–15.

Bretscher, N. (2014). Exploring the quantitative and qualitative gap between expectation and implementation: A survey of English mathematics teachers' uses of ICT. In A. Clark-Wilson, O. Robutti, & N. Sinclair (Eds.), *The mathematics teacher in the digital era: An international perspective on technology-focused professional development* (Vol. 2, pp. 43–70). Dordrecht: Springer.

Burrill, G., Allison, J., Breaux, G., Kastberg, S., Leatham, K., & Sanchez, W. (Eds.). (2002). *Handheld graphing technology in secondary mathematics: Research findings and implications for classroom practice.* Dallas, USA: Texas Instruments.

Burrill, G., & Biehler, R. (2011). Fundamental statistical ideas in the school curriculum and in training teachers. In C. Batanero, G. Burrill, & C. Reading (Eds.), *Teaching statistics in school mathematics-challenges for teaching and teacher education: A joint ICMI/IASE study: The 18th ICMI Study* (pp. 57–69). Dordrecht: Springer.

Clarke, D. M. (1994). Ten key principles from research for the professional development of mathematics teachers. In D. B. Aichele & A. F. Croxford (Eds.), *Professional development for teachers of mathematics* (pp. 37–48). Reston, VA, USA: National Council of Teachers of Mathematics.

Coburn, C. (2003). Rethinking scale: Moving beyond numbers to deep and lasting change. *Educational Researcher, 32*(6), 3–12.

Deci, E. L., & Ryan, R. M. (Hrsg.). (2000). *Handbook of self-determination research.* Rochester: University Press.

DMV, GDM, & MNU. (2008). *Standards für die Lehrerbildung im Fach Mathematik – Empfehlungen von DMV, GDM, MNU.* Retrieved from http://didaktik-der-mathematik. de/pdf/Standards%20Lehrerbildung%20Mathematik.pdf

Drijvers, P., Ball, L., Barzel, B., Heid, M. K., Cao, Y., & Maschietto, M. (2016). *Uses of technology in lower secondary mathematics education: A concise topical survey.* Cham: Springer Open.

Duval, R. (2002). Representation, vision and visualization: Cognitive functions in mathematical thinking: Basic issues for learning. In F. Hitt (Ed.), *Representations and mathematics visualization* (pp. 311–335). Mexico: Cinvestav.

Fishman, B. J., Penuel, W. R., Allen, A.-R., Cheng, B. H., & Sabelli, N. (2013). Design-based implementation research: An emerging model for transforming the relationship of research and practice. In B. J. Fishman & W. R. Penuel (Hrsg.), *National society for the study of education: Vol 112: Design based implementation research* (pp. 136–156). New York: Columbia University Press.

Garet, M. S., Porter, A. C., Desimone, L., Birman, B. F., & Yoon, K. S. (2001). What makes professional development effective? Results from a national sample of teachers. *American Educational Research Journal, 38*, 915–945.

Guskey, T. R. (2000). *Evaluating professional development.* Thousand Oaks, CA: Corwin Press.

Hadjidemetriou, C., & Williams, J. (2002). Children's graphical conceptions. *Research in Mathematics Education, 4*(1), 69–87.

Heid, M. K., & Blume, G. W. (Eds.). (2008). *Research on technology and the teaching and learning of mathematics: Volume 1 research syntheses.* Charlotte, NC: IAP.

Hill, H. C., Loewenberg Ball, D., & Schilling, S. G. (2008). Unpacking pedagogical content knowledge: Conceptualizing and measuring teachers' topic-specific knowledge of students. *Journal for Research in Mathematics Education, 39*(4), 372–400.

Kahneman, D., & Tversky, A. (1972). Subjective probability: A judgment of representativeness. *Cognitive Psychology, 3*, 430–454.

Klinger, M. (2018). *Funktionales Denken beim Übergang von der Funktionenlehre zur Analysis.* Wiesbaden: Springer Spektrum.

KMK (Sekretariat der Ständigen Konferenz der Kultusminister der Länder der Bundesrepublik Deutschland). (Ed.). (2012). *Bildungsstandards im Fach Mathematik für die Allgemeine Hochschulreife (Beschluss der Kultusministerkonferenz vom 8.10.2012).* Köln: Kluwer.

Krainer, K. (2003). Teams, communities and networks. *Journal of Mathematics Teacher Education, 6*, 93–105.

Lam, T. C., & Bengo, P. (2003). A comparison of three retrospective self-reporting methods of measuring change in instructional practice. *American Journal of Evaluation, 24*(1), 65–80.

Lipowsky, F., & Rzejak, D. (2015). Key features of effective professional development programmes for teachers. *Ricercazione, 7*(2), 27–51.

Lorenz, R., Bos, W., Endberg, M., Eickelmann, B., Grafe, S., & Vahrenhold, J. (Hrsg.). (2017). *Schule digital – der Länderindikator 2017. Schulische Medienbildung mit besonderem Fokus auf MINT- Fächern in der Sekundarstufe I im Bundesländervergleich und Trends von 2015 bis 2017.* Münster: Waxmann.

Meyfarth, T. (2006). *Ein computergestütztes Kurskonzept für den Stochastik-Leistungskurs mit kontinuierlicher Verwendung der Software Fathom – Didaktisch kommentierte Unterrichtsmaterialien. Kasseler Online-Schriften zur Didaktik der Stochastik (KaDiSto) Bd. 2.* Kassel: Universität Kassel. Retrieved from http://nbn-resolving.org/urn:nbn:de:he bis:34-2006092214683

Nieszporek, R., & Biehler, R. (2017). Kompetenzzuwachsmessung bei Lehrkräftefortbildungen durch retroperspektive Selbsteinschätzung am Beispiel von "Stochastik kompakt". To be published. *Beiträge zum Mathematikunterricht*.

Penglase, M., & Arnold, S. (1996). The graphics calculator in mathematics education: A critical review of recent research. *Mathematics Education Research Journal*, *8*(1), 58–90.

Pierce, R., & Stacey, K. (2010). Mapping pedagogical opportunities provided by mathematics analysis software. *International Journal of Computers for Mathematical Learning*, *15*(1), 1–20.

Prediger, S., Leuders, T., & Rösken-Winter, B. (2017). Drei-Tetraeder-Modell der gegenstandsspezifischen Professionalisierungsforschung – Fachspezifische Verknüpfung von Design und Forschung. In K. Zierer, et al. (Eds.), *Jahrbuch für Allgemeine Didaktik 2017. Thementeil Allgemeine Didaktik und Lehrer/innenbildung* (pp. 159–177). Baltmannsweiler: Schneider Verlag Hohengehren.

Prediger, S., Leuders, T., & Rösken-Winter, B. (2019). Which research can support PD facilitators? Strategies for content-related PD research in the three-tetrahedron model. *Journal of Mathematics Teacher Education*, 22, 407–425. doi:10.1007/s10857-019-09434-3

Prömmel, A. (2013). *Das GESIM-Konzept – Rekonstruktion von Schülerwissen beim Einstieg in die Stochastik mit Simulationen*. Heidelberg: Springer Spektrum.

Putnam, R. T., & Borko, H. (2000). What do new views of knowledge and thinking have to say about research on teacher learning? *Educational Researcher*, *29*(1), 4–15.

Rösken-Winter, B., Schüler, S., Stahnke, R., & Blömeke, S. (2015). Effective CPD on a large scale: Examining the development of multipliers. *ZDM Mathematics Education*, *47*(1), 13–25.

Rosenbaum, P. R., & Rubin, D. B. (1985). Constructing a control group using multivariate matched sampling methods that incorporate the propensity score. *The American Statistician*, *39*(1), 33–38.

Rösken-Winter, B., & Szczesny, M. (2017). Continuous Professional Development (CPD): Paying attention to requirements and conditions of innovations. In S. Doff & R. Komoss (Eds.), *Making change happen* (pp. 129–140). Switzerland: Springer Nature.

Schacht, F. (2017). Between the conceptual and the signified: How language changes when using dynamic geometry software for construction tasks. *Digital Experiences in Mathematics Education*, Online first. doi:10.1007/s40751-017-0037-9

Schoen, D. A. (1983). *The reflective practitioner: How professionals think in action*. London & New York, NY: Routledge Taylor & Francis Group.

Sedlmeier, P., & Gigerenzer, G. (1997). Intuitions about sample size: The empirical law of large numbers. *Journal of Behavioral Decision Making*, *10*(1), 33–51.

Swan, M. (Ed.). (1985). *The language of functions and graphs: An examination module for secondary schools*. Nottingham: Shell Centre for Mathematical Education.

Sztajn, P., Borko, H., & Smith, Th. (2017). Research on mathematics professional development. In J. Cai (Ed.), *Compendium for research in mathematics education* (pp. 793–823). Reston, VA, USA: National Council of Teachers of Mathematics.

Tall, D., & Vinner, S. (1981). Concept image and concept definition in mathematics with particular reference to limits and continuity. *Educational Studies in Mathematics*, *12*(2), 151–169.

Thurm, D. (2017). Psychometric evaluation of a questionnaire measuring teachers beliefs regarding teaching with technology. In Kaur, et al. (Hrsg.), *Proceedings of the 41st Conference of the International Group for the Psychology of Mathematics Education* (pp. 265–272). Bd. 4. Singapore: PME.

Thurm, D. (2020). *Digitale Werkzeuge im Mathematikunterricht integrieren – Zur Rolle von Lehrerüberzeugungen und der Wirksamkeit von Fortbildungen*. Wiesbaden: Springer.

Thurm, D. & Barzel, B. (2020). Effects of a professional development program on teachers' beliefs, self-efficacy and practices. *ZDM Mathematics Education.* https://doi.org/10.1007/s11858-020- 01158-6

Thurm, D., Klinger, M., Barzel, B., & Rögler, P. (2017). Überzeugungen zum Technologieeinsatz im Mathematikunterricht: Entwicklung eines Messinstruments für Lehramtsstudierende und Lehrkräfte. *Mathematica didactica*, 40 (1), pp. 19-35.

Timperley, H., Wilson, A., Barrar, H., & Fung, I. (2007). *Teacher professional learning and development: Best evidence synthesis iteration.* Wellington: Ministry of Education.

Tversky, A., & Kahneman, D. (1971). Belief in the law of small numbers. *Psychological Bulletin, 76*(2), 105–110.

van den Akker, J., Gravemeijer, K., McKenney, S., & Nieveen, N. (Eds.). (2006). *Educational design research.* London: Routledge.

vom Hofe, R., & Blum, W. (2016). "Grundvorstellungen" as a category of subject-matter didactics. *Journal für Mathematik-Didaktik, 37*(suppl 1), 225–254.

Wassner, C., Biehler, R., Schweynoch, S., & Martignon, L. (2004). Authentisches Bewerten und Urteilen unter Unsicherheit – Arbeitsmaterialien und didaktische Kommentare für den Themenbereich "Bayessche Regel". In C. Wassner (Ed.), *Förderung Bayesianischen Denkens: kognitionspsychologische Grundlagen und didaktische Analysen.* Hildesheim: Franzbecker.

Wassong, T., & Biehler, R. (2010, July). A model for teacher knowledge as a basis for online courses for professional development of statistics teachers. In C. Reading (Ed.), *Data and context in statistics education: Towards an evidence-based society: Proceedings of the Eighth International Conference on Teaching Statistics, ICOTS8, Ljubljana, Slovenia.* Voorburg, The Netherlands: International Statistical Institute. Retrieved from www.iase-web.org

INDEX

For Product Safety Concerns and Information please contact our EU
representative GPSR@taylorandfrancis.com
Taylor & Francis Verlag GmbH, Kaufingerstraße 24, 80331 München, Germany